THE UNIVERSE ISN'T JUST A BUNCH OF ROCKS

PAUL E. WILLIAMS

EDITED BY CAROL M. WILLIAMS

THE UNIVERSE ISN'T JUST A BUNCH OF ROCKS

The Universe Isn't Just a Bunch of Rocks
© 2012 Paul E. Williams

All rights reserved. No part of this book may be reproduced or transmitted in any form or by any electronic or mechanical means, including photocopying, recording, or by any information storage, or retrieval system known or in the future without written permission of publisher and/or author except as permitted by law.

Library of Congress Cataloging-in-Publication Data

The Universe Isn't Just a Bunch of Rocks

Williams, Paul E., 1929 -

ISBN-13: 978-1477593554

ISBN-10: 1477593551

Edited by Carol M. Williams
Cover design by Wendy L. Williams

First published June 2012
Printed in the United States of America

Dedication

To Rosemary, my perfect wife of 54 years who succumbed to Alzheimer's disease two years ago after having it for ten years.

Rosemary never stopped lovingly supporting me in any endeavor and raised our five children as model citizens who have her stamp of morality and confidence on each of them.

And -
To God, who picked out Rosemary and handed her to me and who has helped me constantly with this book and with my life.

CONTENTS

FOREWORD 1

SYNOPSIS 3

INTRODUCTION 5

1   WHAT IS THE UNIVERSE? 9

2   DID THE UNIVERSE ALWAYS EXIST? 25

3   INFINITY IS IMPOSSIBLE 27

4   THE ORIGIN OF THE UNIVERSE 47

5   THE NON-UNIVERSE 53

6   CAN YOU DESCRIBE THE CREATOR OF THE UNIVERSE? 59

7   UNIVERSE PHILOSOPHY 71

8   THE OBJECT-ORIENTED UNIVERSE 77

9   CAN EVOLUTION TOTALLY EXPLAIN THE UNIVERSE MAKE UP? 81

10   WHAT ABOUT THE UNIVERSE EQUATIONS? 95

11   PAUSE FOR A MOMENT, JUST FOR EMILY 97

12	ARE TIME MACHINES POSSIBLE IN THE UNIVERSE? 101
13	THE WILDLY FANTASTIC FUTURE BECAUSE OF COMPUTERS AND ADVANCED TECHNOLOGY 121
14	GETTING THERE FASTER THAN AT "LIGHT SPEED" 137
15	THE VIRTUAL WORLD OF THE FUTURE 143
16	OUR SPACE TRAVEL IN THE UNIVERSE 155
17	CAN WE CONTACT OTHER PEOPLE IN THE UNIVERSE? 159
18	THE UNIVERSE - THE FUTURE OF HUMANS 175
19	OUR RESPONSIBILITY TO LIMIT POPULATION, IN THE FUTURE 179
20	WHAT ABOUT "MIRACLES" 185
21	ABOUT HEAVEN 191
22	A SIMPLE CONCLUSION 195

ACKNOWLEDGEMENTS 197

ABOUT THE AUTHOR 201

Foreword

In this book, an attempt is made to put things in a language so that people of different educational backgrounds may understand. While it, necessarily contains some technical information, a serious attempt is made to present things as simply as possible.

Also, an attempt is made to present things from several viewpoints. I think that it is especially important to take a philosophical view as well as a technical one - and, maybe, with a little humor at times, as well.

Finally, I am well aware that some will be offended by some conclusions, but this is necessary when presenting reasoned, logical information which is, sometimes in direct conflict with some established concepts. These old concepts may sometimes be "blind faith" tenets that may have been stamped and ingrained and enforced, en-pushed and "sealed in concrete" for many years.

In evaluating possibilities of the Universe (and other things, as well), I believe a correct rule is: **When it's true, it's true** and "Let the devil take the hindmost."

THE UNIVERSE ISN'T JUST A BUNCH OF ROCKS

Synopsis

The Universe is made up of a lot more than just matter, including: Matter, Energy, Waves, Electrical, Space, Time, Rules/Laws/Equations, Souls?, Links To God? - *Proven & Reasoned Conclusion*

The "Rules/Laws" are, in many ways, more important than "Matter" in the universe - *Reasoned Conclusion*

The universe didn't always exist, since it would have been for an infinity of time - *Proven*

There is no Infinity (or "Paradoxes" or "Everything"). This is against the definition - *Proven*

Time, space, matter and the "rules" (gravity, time/velocity, atom behavior, sub particle actions, etc.) are part of the Universe and they started when the universe started - *Proven*

The universe didn't start by itself. There must have been a master planner and creator we will call "God" - *Proven*

This "God" is *really* something *extra capable* (not an old man with whiskers in a rocking chair or a *vindictive/shallow/changes his mind/gets mad/etc.* "guy") - *Proven*

God can't do *everything*, this is against the definition of *everything* - *Proven*

Evolution is nature's/God's method of development - *Proven*

Everything wasn't done by evolution alone using only chance-with-natural-selection following - *Reasoned Conclusion*

THE UNIVERSE ISN'T JUST A BUNCH OF ROCKS

We won't be able to get/receive radio signals from/with aliens from other solar systems, or have their actual bodies here on earth, because of the "400-year window" mismatch - *Reasoned Conclusion*

There aren't/won't ever be any "time machines" since there is a basic impossibility in two-way time travel. Also, nobody from the future or from other solar systems has communicated with us - *Reasoned Conclusion*

There aren't miracles (past or future) which *violate* nature's laws. After all, God set up the rules and wouldn't/couldn't/didn't need to violate them - *Reasoned Conclusion*

Miracles can be performed due to "adjustment of probabilities" or "advanced technology" - *Reasoned Conclusion*

Future (100 to 300 years) computers and related development will allow us to make almost anything happen, absolutely "like magic" - *Reasoned Conclusion by Simple Extrapolation*

Whatever the afterlife for "souls," *heaven* has no physical matter; therefore there is no *heaven* as envisioned with "physical images" or "talking with words" or "St. Peter at the gate." We don't know about the exact future for souls, after all, *nobody has ever come back* - *Reasoned Conclusion*

If actions are always deterministic and not by free will, then people's actions don't make any difference but, if free will does exist, then it does make a difference. So, to be on the safe side, it's wise to assume free will - *Reasoned Conclusion*

Introduction

To start with...

When the Universe is considered as many, many billions of stars, gaseous clouds, space, time, and the like, it is seen as a fantastic thing. However, this pales and becomes relatively insignificant when considering the rules/laws that hold the Universe together and make everything work. These rules/laws are really, really, really something!

But, then, when it is considered that the Universe couldn't have always existed, or it would have existed for an infinity of time periods (impossible) and that, if it thus started, it couldn't have just started by itself, the creative force that did this is R-E-E-A-L-L-Y something!

Think about it.

Did the Universe always exist?

Our moon orbits the earth and is in "balance" with the rest of our solar system. It has effects on the earth's tides and reflects the sun for our nighttime romantic strolls.

What if these things stayed the same but the moon was made of soft, green cheese? Would it melt in response to the sun's heat? Would mice be on the moon to eat the green cheese and cats reside there to eat the mice?

The above is nonsense since the moon *isn't* made of soft green cheese. Any soft green cheese object wouldn't orbit the same, etc. When you pose a question like this, you are just playing games, not dealing with real things.

THE UNIVERSE ISN'T JUST A BUNCH OF ROCKS

One of my favorite subjects is the erroneous treatment of infinity in a similar manner to the description of the moon as being soft green cheese. It is often treated this way by many renowned physicists, philosophers and mathematicians.

The meaning and definition of infinity is that it is endless. No count or quantity can ever reach infinity since there is always "at least one more."

Still, many, otherwise competent, experts use infinity as though it could be actually reached. Dr. Brian Greene, a renowned physicist talks of "an infinite universe" where everything (of course) is repeated, so there is two of each of us, etc. The great mathematician, Hilbert's "Grand Hotel" paradox is a well-known story coming up with all kinds of, what I call, "fairy tales" using an assumption of infinity being reached. The well-known physicist, Stephen Hawking, even talks of an infinite number of universes.

Since the definition of infinity is that it can't be reached (there is always one more), you can't (without first having a couple of martinis) reasonably state what would happen if it really were reached.

In a like manner, you ***can't theorize that the Universe always existed*** since it then would have existed for an infinite number of time periods.

The Universe's makeup

The origin of the Universe is a question that multitudes of people have been interested in for centuries.

Usually, some theory such as the "Big Bang" theory is used to start explaining the origin. The basic assumption is that the Universe consists of matter and energy. The question is, how do you explain the origin of these?

INTRODUCTION

While the Universe does contain matter and energy, these aren't the "whole thing" at all. There are a number of known components of the Universe and, undoubtedly, a *large* number of unknown ones too. We know about how it is possible to convert matter into energy (nuclear conversion). When we think about the origin of matter and energy, an even more intriguing question is: "Where did all of the *rules/laws* come from?"

While the creation of matter and energy is interesting: *How* or *Who* set up and started the *Rules*?

Many people have searched for a "unified theory" of the Universe. Einstein did so. Steven Hawking has made it a real life goal to come up with a "unified theory" that would explain the makeup of the Universe. This means a theory that would explain, not only space, matter, gravity and time but, also, the makeup of each of these. Perhaps, someday, we might have a unified theory which builds on sub-microscopic layers of matter. This might even go to the level of "string theory" where electrons, protons and nuclei are made up of "one dimensional" energy "strings."

While there might be sets of theories and equations that could somehow tie many of these things together, there is still more to the rules than just the elements which make up matter; even more than those found after linking electromagnetic forces, etc.

There are also the rules or laws such as "An object remains in motion until something changes it" and "Bernoulli's laws of fluid dynamics." There are other rules such as the use of the number Pi, which is used in many equations. And, how about all of the equations in physics and geometry! It will be a long time before we will ever be able to explain all of these and, I doubt if we will find all of them "unified" to one simple "set."

Also, I don't see the Universe as having been set up randomly by chance or by chance with evolutionary development or on a Sunday afternoon by a school child.

THE UNIVERSE ISN'T JUST A BUNCH OF ROCKS

So, what do we know?

We know that Infinity can't be reached, period.

We know that the makeup of the Universe contains a number of known and many unknown items

The actual creation of matter and energy is an important one. It is one that we couldn't just perform as a third grader. However, the **creation** of all the **rules** of the Universe is **really quite an achievement**. It stands alone, making just the creation of *matter* as relatively simple, by comparison.

What part do people play in the Universe?

Were people created on earth to satisfy some grand design? Or, were people created accidentally, and evolved through the evolutionary process? We can't prove to a 100% probability, the answers to this question but we can find some clues.

Are we the only people in the Universe? Can we hope to come in contact with other people if they exist? Here, the key question relates to the relative evolutionary periods of our and other races on other planets - if they exist.

We need to examine the logic of evolutionary development of humans in order to determine a lot of the answers about the possibility of other people in the Universe.

1 What is the Universe?

How much is a billion (1,000,000,000)?

Before we start discussing billions of years, galaxies, etc. let's mention that a billion is quite a lot.

I once started to count to a billion to see how much it was. I was counting - one hundred, eighty two thousand, three hundred twenty two. Then, one hundred, eighty two thousand, three hundred twenty three....

I thought, this is going pretty slowly so I started to do some calculations. If I count really fast, I can count at the rate of maybe one count in three seconds for 24 hours a day, seven days a week. If I started at age two on my second birthday, then on my 102nd birthday, 100 hundred years later, I would finally count to one billion.

OK, it's quite a lot.

The size of the physical Universe

The age of the Universe is estimated at about 13.7-billion years. Our galaxy, the Milky Way galaxy is estimated to be 13.2-billion-years old, nearly as old as the Universe.

The earth is in the Milky Way galaxy, in our solar system. We have the planets Mercury, Venus, Earth, Mars, Jupiter, Saturn, Uranus, and Neptune, plus the dwarf-planet Pluto in our solar system. We only have one star in our solar system, the Sun. The nearest star to the Sun is Proxima Centauri, about 4.3 light-years away.

There are 200- to- 400-billion stars in the Milky Way galaxy.

THE UNIVERSE ISN'T JUST A BUNCH OF ROCKS

Light travels at about 186,000 miles per second. A light year is about 5,878,625,373,184 miles. That's 5.9 **trillion!** For you computer people, if you had a six-terabyte memory, that's one byte of memory for each mile in a light year.

There are estimated to be at least 115 galaxies within 12-million light-years away. Each galaxy may be between thousands and, maybe, millions of light years in diameter.

It is estimated that there are 100- to- 400-billion galaxies in the Universe. Many more are being discovered all of the time.

The diameter of the observable Universe is about 93-billion light years. The actual Universe size is estimated at somewhere between 78-billion light years and 156-billion light years.

By contrast, the distance to the Earth's moon is, on average, slightly more than one *light-second* away.

It is interesting that the estimated radius of the Universe is about 39- to- 78-billion light years. This would seem to indicate that objects in distant regions of the Universe may be getting further away from each other at speeds greater than the speed of light.

However, this isn't necessarily true. The fact is that this size of the Universe is due to the expansion of the space that separates these objects rather than movement in the conventional sense of the word.

WHAT IS THE UNIVERSE?

Universe makeup

The following is a chart showing many of the components of the Universe.

Universe Makeup

Matter/ Energy	Buildings/ Paintings Organics
Waves	Electromagnetic (including Light Waves Microwaves Etc.)
Electrical	Electricity Magnetic Fields
Space	
Time	
Rules/Laws	Gravity, Motion Laws, Etc.
Souls?	
Links To God?	
Black Holes, Dark Energy?	
Other?	

→ Compounds→Elements→Molecules→Atoms→Neutrons→Quarks→Strings

THE UNIVERSE ISN'T JUST A BUNCH OF ROCKS

Matter/Energy

Strings, Ions, Electrons, Protons, Atoms, Molecules --- all the way to --- Galaxies.

It's a pretty complex organization of stuff, this matter.

The little, tiny atom things are made up of their components. They have protons, neutrons and little electrons whirling around like all get out. Scientists call the even smaller things, *elementary* and *composite* particles. *Elementary* particles are those that don't break down any further, while *composite* particles are made up of even smaller particles.

I would write about the things that make up the really little parts and why the parts do the things that they do, but some of it is argued about, a lot.

Scientists get into really hot arguments when they go to the level of quantum physics. One basic problem that they have is trying to look at the little things. They put them under the microscope and try to shine a light on them so that they can see them. Now, when you shine a light on something, you say that "you hit it with a light." When a quantum person does that, he says that he "hits it with a photon." By the way, according to scientists, starting with Al Einstein, light, which is shining on things, is really a bunch of photons which are the actual little "make things shine" things. They don't weigh anything since they don't have any mass but they have some, major energy, which does affect things (little things and big things like solar sails used for space propulsion).

It is really interesting when you bang up against a really little particle with a photon or two, since the photon wrecks the little particle so that they aren't the thing that you were trying to look at. Boy is that frustrating!

WHAT IS THE UNIVERSE?

When the scientists can't see the really little things, they use mathematical equations to represent what they think the little things are doing. They really have a lot of equations. (Sometimes, I think that some scientists like equations better than sex and that's a lot of liking.)

Anyway, partly because the scientists can't easily see the sub-atomic particles and partly because of their calculations, some say that things keep disappearing and reappearing - according to probabilities, not fixed rules.

(If you wonder how many atoms there are, it has been estimated that, in our Earth, there are $133*10^{48}$ atoms. That's 133,000, ---- with 45 more 0s.)

Elements are made up of atoms. There are around 92 natural atoms/elements but more have been created by scientists.

Molecules use combinations of elements to make up the materials of the Universe.

Combinations of these materials then are used to make up the many galaxies of the Universe.

On Earth, and possibly in other places in the Universe, organic matter is added to inorganic matter.

Organic matter is any part of any substance which once had life. That is, matter that is or has come from, plants or animals.

THE UNIVERSE ISN'T JUST A BUNCH OF ROCKS

Electromagnetic Waves

Some frequencies and wave lengths are shown:

Type	Frequency (Hz)	Wavelength (m)
Radio waves	$< 3 \times 10^8$	> 1
Microwaves	$3 \times 10^8 - 7.5 \times 10^{11}$	$0.004 - 1$
Infrared Light	$7.5 \times 10^{11} - 4.3 \times 10^{14}$	$7 \times 10^{-7} - 0.004$
Visible Light	$4.3 \times 10^{14} - 7.5 \times 10^{14}$	$4 \times 10^{-7} - 7 \times 10^{-7}$
Ultraviolet	$7.5 \times 10^{14} - 6 \times 10^{15}$	$5 \times 10^{-9} - 4 \times 10^{-7}$
X-rays	$6 \times 10^{15} - 5 \times 10^{19}$	$6 \times 10^{-12} - 5 \times 10^{-9}$
Gamma Rays	$> 5 \times 10^{19}$	$< 6 \times 10^{-12}$

We know about some of the principles involved with these waves but, again, we don't know anything about the **why.**

We do know some things about the **rules** of these waves, primarily light waves, since we can see them all over the Universe.

We know about the speed of light waves. (This is assumed to be the same for all of these kinds of waves.) In a vacuum and with no velocity or gravity forces considered, this is about 186,000 miles per second, 700,000,000 miles per hour or 300,000,000 meters per second. (Exactly 299,792,458 m/s.) Electricity travels slower through mediums such as copper wire, about 40% to 70% in twisted pairs of copper wires. It travels at 40% to 95% in all mediums, depending on the medium.

We also know that this speed is an absolute constant - it can't go any faster even if the light source is moving at a fast clip in the same direction as the propagation of the light. Einstein theorized that this was caused through time "dilation" which means variations to time between different clocks.

WHAT IS THE UNIVERSE?

We also know that time slows with increased gravity and with velocity. In the Global Positioning System (GPS), adjustments are necessary to compensate for this. Time dilation due to velocity formula is: **$t'=t/\sqrt{(1-(v^2/c^2))}$,** where t and t' are the two observers times, v is the velocity and c is the speed of light. The difference in time for the clock in the satellite and on the Earth's surface, for 1 second is given by: **$t'=1/\sqrt{(1-(2gm/rc^2))} - 1/\sqrt{(1-(2gm/(r+a)c^2))}$**, where g is the gravitational constant, m is the mass of the Earth, r is the radius, a is the altitude of the GPS and c is the speed of light.

This turns out to be about 45 microseconds per day due to gravity and about -7 microseconds per day due to relative velocity for a net of about 38 microseconds per day adjustment.

Light effects - on magnetic fields and gravity

The forcing effects of light on an object are quite small but still there. While scientists agree that light (and other waves) don't have any mass to them, light does exert a force on an object. They say that light doesn't have "mass" but does have "momentum." They then call the "things in light" that cause this effect - "photons." Of course, nobody knows what a photon really is. Light also is known to cause a very small magnetic field and recent discoveries have suggested that this may be multiplied by millions of times over previous estimates to allow better, future generation of electricity from the Sun.

Also, experiments have shown that light apparently bends due to gravity. The question is whether it bends or rather, whether space, through which it is travelling, is warped.

THE UNIVERSE ISN'T JUST A BUNCH OF ROCKS

Electricity and magnetic fields

We might separate electricity from electromagnetic waves since there are such differences in applications. While we use magnets and electric motors in almost every area of our lives, we don't know much about electricity or magnetic fields. We don't even know how fast electricity travels - by itself. We can measure its speed in each medium such as in copper wires but how does electricity relate to light waves?

We know about electrically created magnets and motors. We know there are magnetic fields because we can track them with iron filings, etc. However, we don't know much about any of these things except the results that we can see - like magnetic attraction and workhorse motors.

A simple electromagnet consists of a coil of insulated wire wrapped around an iron core.

Motors create almost all of the movement in appliances in our homes as well as in many thousands of applications elsewhere.

But we still don't know much of anything about **why** electromagnetic fields work.

Space

Space may or may not exist without matter or electromagnetic waves, but it certainly does exist when matter is present. Be sure to remember that space is not, just "nothing." It takes time to go across a mile of space, etc. Space seems to be part of the Grand Equations involving space, time, matter and energy.

WHAT IS THE UNIVERSE?

Time

Many people describe time as "a measure of change" (of matter). Using this description, without the Universe, if there is no matter, there is no time.

It does appear that measuring change in matter provides a measure of time. Time *could*, in theory, exist alone, but it's a safe assumption that time needs matter and the other parts of the Universe.

Reason, logic, and common sense seem to indicate that all of the kinds of things that make up the Universe are united, work together, and were probably formed together at the start of the Universe.

Rules/laws

What about the rules (laws), those that make electrons fly around atoms? What about gravity? What about the "clumping together of atoms" rules to make molecules? What about the limited number of basic elements in the periodic table? What about time dilation with velocity and gravity and Black Hole phenomenon? When did the rule that states that "an object in motion will remain in motion until some force changes it" get created?

This list goes on and on and there is even the possibility, considering quantum mechanics questions, that the rules and energy may be all that there is in making up the matter and energy. That is to say that matter may be just the "rules of how energy acts to ′make up′ matter." In fact, energy might just be some more rules, too.

A good motto is: **"Rules Rule."**

THE UNIVERSE ISN'T JUST A BUNCH OF ROCKS

Souls

We know that people are at the top of the evolutionary chain of organics. We feel that people have some special place in the Universe. According to different religious beliefs, we may believe that a portion of our makeup may survive after death. Our bodies consist of our various parts such as arms, legs, etc. and we have brains with memories. However, so do other animals. Most of us, feel that we humans are special. Part of this is because of our superior reasoning capabilities allowing us to conquer lower animals in the chain. Also, because we can do all kinds of things like write books, fly in airplanes and carouse in nightclubs.

We also know that most of our bodies "go" at the time of death. This seems to include our various parts such as arms, legs, stomachs and hair (which usually went long before death for us men). It makes sense to assume that our brains go too. And, along with that we have to assume that our primary memory storage devices go "kaput" after we "kick the bucket."

The question is: What is there that "keeps on ticking" after our hearts stop doing just that?

Whatever this is that survives, we usually call the "soul." If this is the surviving thing, then it must have been there while we were alive. Therefore, we can include it in the "Universe inventory."

WHAT IS THE UNIVERSE?

Some "Just for Consideration Ideas" about souls

Was each of our individual souls created when we were conceived as humans?

Or, was each soul taken from a pile and meted out at that time?

Or, was our soul recycled, having been used before? This fits, in some way, with the idea of reincarnation. The idea is that our soul might develop for more than one - perhaps - many generations. Reincarnation does answer some of the questions of fairness relating to the different chances that one person had over another. For example, if a child was born and lived for one day or one week or one year, he didn't have the same chance to develop his soul as another person. In a like manner, a person born blind, deaf and mute hasn't had the same chances to "work with his soul" as the mountain climber, explorer turned missionary to the disadvantaged, or the Monk who finally made Pope." (A fairly long shot idea.)

Or, was each soul a part of an overall "mass soul" of some kind, before they "split off?"

I guess it's even possible that we could start out without a soul and have it develop as we go along. Oops - forget that one or I'm in big trouble with lots of people, including the right-to-life people.

Does our soul go to Heaven (or someplace else, maybe) as an individual soul? Or, does it go as a composite part of things?

To tell you the truth, I don't think that it matters much about this question as long as we do have a soul and don't let it go to the devil.

So, if I were you, I wouldn't sit up nights worrying too much over this one.

Just keep on eating lots of Soul food* and you'll be OK, I think.

THE UNIVERSE ISN'T JUST A BUNCH OF ROCKS

*Soul food is an American cuisine, a selection of foods, and is the traditional cuisine of Black Americans of the southern United States and among many black communities in other areas of the United States. It is closely related to Southern cuisine of the United States.

Links to God

I can show that the Universe didn't always exist. If it started, it makes sense that there may have been a creative force that started it.

Let's call this creative force - God. (Note: I believe that most religious beliefs usually follow this line of reasoning although many of them add a lot more in addition to just that starting point.)

This creative force existed before the Universe existed and therefore wasn't part of the Universe. So, God isn't classed as part of the Universe.

However, most people in most religions pray to God. Since we are in the Universe, we need a link to get there. That's why I include, in the Universe contents, a "link to God." Voila!

Note: If you insist that God *is part of the Universe*, then just substitute God for this link.

Black Holes, Dark Energy and Other Items of Interest

There are lots of other universe contents. Some are proven, like black holes. Some are fairly solid theories, but not really proven, like Dark Energy. Others are just conjectures, some by wild scientists.

Probably the most important "others" are the ones that we don't know of, at present.

WHAT IS THE UNIVERSE?

Black Holes

A black hole is a place where gravity is so great that nothing, not even light itself, can escape the pull of gravity there. As a result, it cannot be seen directly. It can, however, be seen by its effect on nearby stars, which will orbit around it. It may also be seen because of what is called gravitational lensing. That's a situation where a star or other light-emitting object is "behind" the black hole and the light bends around the black hole because of the gravitational effects of the black hole.

So, we can't see them but we know that black holes exist. We know how they are born, where they occur, and why they exist in different sizes. We even know what would happen if you fell into one.

The theory of general relativity predicts that a sufficiently compact mass will deform space-time to form a black hole. Around a black hole there is a mathematically defined surface called an event horizon that marks the point of no return. It is called "black" because it absorbs all the light that hits the horizon, reflecting nothing. It is also theoretically possible to observe the black hole by virtue of the action at its event horizon. Light and higher energy electromagnetic radiation will be generated there by particles of matter being accelerated into the black hole itself. But you'd need to be close to see this.

Imagine that we are on a spaceship near a black hole. We drop a clock into the black hole and compare its time to that of our onboard clock. The falling clock runs progressively slower. It never crosses the event horizon, but stays frozen there in space and time. The falling clock also becomes continuously redder, since its light loses energy as it escapes from the black hole's vicinity.

By contrast, if we were falling *with* the clock, time, to us, would appear to behave normally.

THE UNIVERSE ISN'T JUST A BUNCH OF ROCKS

Dark energy and dark matter

Dark energy is a **hypothetical** form of energy that permeates space and exerts a negative pressure, which would have gravitational effects to account for the differences between the theoretical and observational results of gravitational effects on visible matter. Dark Energy increases the rate of expansion of the universe.

Dark energy is the most accepted theory to explain recent observations that the universe appears to be expanding at an accelerating rate. In the standard model of cosmology, dark energy currently accounts for 73% of the total mass-energy of the universe.

Dark energy is not directly observed, but rather inferred from observations of gravitational interactions between astronomical objects, along with *dark matter*.

Before physicists knew about dark energy, a cosmological constant was a feature of Einstein's original general relativity equations that caused the universe to be static. When it was realized that the universe was expanding, the assumption was that the cosmological constant had a value of zero - an assumption that remained dominant among physicists and cosmologists for many years.

In 1998, scientists attempting to measure the deceleration of the universe's expansion measured not only a deceleration, but an unexpected *acceleration*.

Further evidence since 1998 has continued to support this finding that distant regions of the universe are actually speeding up with respect to each other. Instead of a steady expansion, or a slowing expansion, the expansion rate is getting faster, which means that Einstein's original cosmological constant prediction manifests in today's theories in the form of dark energy.

WHAT IS THE UNIVERSE?

Wormholes

A wormhole is a **hypothetical** interconnection between two regions of space time.

In 1935, Albert Einstein and Nathan Rosen realized that general relativity allows the existence of "bridges," originally called Einstein-Rosen bridges but now known as wormholes. These space-time tubes act as shortcuts connecting distant regions of space-time. By journeying through a wormhole, you could travel between the two regions faster than a beam of light would be able to if it moved through normal space-time. Until recently, theorists believed that wormholes could exist for only an instant of time, and anyone trying to pass through would run into a singularity. But more recent calculations show that a truly advanced civilization might be able to make wormholes work. By using something physicists call "exotic matter," which has a negative energy, the civilization could prevent a wormhole from collapsing on itself. Perhaps, some day in the far future, this could be realized.

Other

In physics, the graviton is a hypothetical elementary particle that mediates the force of gravitation in the framework of quantum field theory. If it exists, the graviton must not have any mass.

A tachyon is a hypothetical subatomic particle, in the field of quantum field theory that moves faster than light and that could be used to send signals faster than light.

There are many other hypothetical particles. Remember that there are, almost surely, other as yet unknown elements which are part of the Universe and fit into the Universe equations. We don't know what they are or what they do but we absolutely must leave room in our reasoning processes for them.

THE UNIVERSE ISN'T JUST A BUNCH OF ROCKS

Perhaps there are "rules" that we don't yet know about or "other things" out there. Remember, we can't see very far out into the Universe since light only travels at about 186,000 miles per second and many, many galaxies are lots and lots of light *years* away. The nearest star to our Sun is about 4.2 light years away. The nearest galaxy to ours is the Canis Major Dwarf galaxy, about 25,000 light years away. We won't get *pictures* from them *in their current states* for a long, long time.

Since the distance to the "edge" of the universe is estimated to be 40 to 80 billion light years away, who knows what we might be missing?

2 Did the Universe Always Exist?

Some argue that the Universe has always existed. This theory doesn't have to wrestle with the question of how things got started in such a profound and orderly manner since "It just always was".

If the Universe has always existed, it would have existed for an infinite number of time periods.

In the next section of this book, the argument is made, using what seems to be unquestionable logic, that reaching *Infinity* is absolutely **impossible in any way** and, therefore, the Universe couldn't have always existed.

It's just that simple!

.

THE UNIVERSE ISN'T JUST A BUNCH OF ROCKS

3 Infinity Is Impossible

To start with …

If you ask my 12-year-old granddaughter, Julie Miller, about Infinity, she will immediately say "Infinity's impossible! Infinity isn't real! It's just a definition of something that's impossible!

Yet, many competent, intelligent people have stated that Infinity in some way was *possible*. These include:

David Hilbert, the German mathematician, who presented the famous "Grand Hotel" story as though it was logical - while the story, itself is just a simple, logical scam.

Georg Cantor, a world famous mathematician renowned for his proofs of "Infinite Sets" - which can't actually exist at all, even in theory or something.

Stephen Hawking, a world famous physicist, adheres to the theory that there are actually many universes. Then, Steven Hawking goes on to theorize that there may be an infinite number of these universes - an impossibility.

Brian Greene, a well-known and competent theoretical physicist at Columbia University, states unequivocally in his book that there is an infinite number of universes. That there are many of these universes which exist, can't be immediately disproved, but to say that there is an infinite number of them - how can the definition of infinity be directly violated?

THE UNIVERSE ISN'T JUST A BUNCH OF ROCKS

In his talks, he states that the number of universes is infinite. He calls this a "multiverse". He then continues by listing some of the impossibilities which would, of course, be encountered when dealing with an infinite number of universes. He makes statements like, "There must be an infinite number of you and me". Clearly, since it is impossible to reach infinity - by definition - how can this be?

Even Aristotle, who agreed that it wasn't possible to reach infinity, used the term "potential infinity" when he apparently meant that something went on without stopping. While it **is** correct to say that something will never end, it seems misleading to include the word "infinity" in such a statement.

There have been many people who have fallen into the trap of stating that infinity can somehow, sort of, be reached.

Is my granddaughter smarter than these people? She is quite smart but these people are very intelligent and quite capable.

I think that these people want to believe in fantasies. They want to believe in *Ali Baba and the Forty Thieves* and the magic *Open Sesame* idea that was described in this fantasy story. There is also the problem that when you are used to following mathematics you sometimes *leave reasoning, at a top level, behind*. Brian Greene has come up with some very interesting - and maybe possible - theories. However, how do you resolve the ambiguity of reaching infinity when the definition is that it can't be reached?

Why is the discussion about Infinity important?

For hundreds, if not thousands, of years people have been fascinated about the subject of infinity. Many, including prominent philosophers and mathematicians have assumed that infinity could somehow, "sort of," be reached. They have used all kinds of words to make it seem as though it was possible. Some (including Aristotle) have talked about "potential" or "theoretical" infinity.

INFINITY IS IMPOSSIBLE

It is **absolutely essential** for us to understand and be clear about infinity for these reasons:

1. Showing that infinity **cannot** be reached shows that the Universe couldn't have existed forever. If it had, the count of time periods would have reached infinity - and you can't do that.

2. Many people talk about God or a supreme being capable of doing anything. Well, this can't *be* because *anything* is another version of infinity; neither of which can be reached for the same reasons. It isn't because God is not powerful enough. It just defies the definition.

Background of infinity discussion

To start with, let's examine three kinds of things: Beliefs, Postulates and Definitions.

Beliefs

Beliefs are those things to which you apply a percentage of correctness. You then use these percentages whenever a situation arises where the theory is germane. There also may be a *degree* applied to the belief.

For example, you may believe that the biblical story of Adam and Eve is basically true. You might give a percentage of 90% to it being true (or 10% or 50%, etc.).

At the same time, you may consider the *degree* to which it is true. This could vary from the story being:

1. Completely true

2. Almost true with a little added for emphasis.

3. Based on a true story but with a lot of embellishment.

4. A fictional story applying the basic meaning.

5. Completely invented.

Postulates

A postulate is a theory for which you or others have offered proofs. When the proofs are very good, they are then used in making decisions. Even though the proofs seem airtight, flaws may be found later which will show that the postulates are not proven after all.

At the same time, you may consider the *degree* to which it is true.

Definitions

A definition is something which *is* because someone has defined it. It is not a question of it being true - it simply is because it was defined that way. It can never be found that something can go in opposition to a definition since it simply **is**. Tricks can be used to make it sound as though the definition is overturned; but they are only tricks, not in any way valid.

For example, if you use the number system of "base 10," 2 plus 2 equal 4. This can't be validly disputed. Even if you come up with a trick to show that some 2 plus 2 was something else, it isn't - because this is a *definition*.

Another example: Black and white colors are two different colors (in our normal way of understanding). You can't try to present a case that states what would be true if black and white were the same color.

I have a very intelligent, technically competent friend who I asked some time ago to critique my theories about infinity. He stated that perhaps there was something that we didn't know about now, which we would find out about and might explain how we could somehow reach infinity.

INFINITY IS IMPOSSIBLE

Well, we can't find out anything which will allow us to violate a definition. That would be in contradiction of the definition.

The definition of infinity

Most people will agree with the standard definition of infinity.

A dictionary definition of infinity is:

The quality of being infinite. Endless or unlimited space, time, distance, quantity, etc. An indefinitely large number or amount.

The key word is "endless."

This means that whatever quantity or number is reached, there is always more. It means that the series of numbers or counts **never ends**.

This is the definition; but many, many people have tried to use it as though a series **does end**. Sometimes they say it is a *concept* of infinity or that it is *conceptual* in nature. Well, it isn't a concept and there isn't anything conceptual about something which doesn't exist and is only a statement of impossibility.

So that should be the end of that!

What infinity is not

Infinity isn't an "it." It isn't a "thing." In fact, it "isn't." The word "infinity" is simply a statement of impossibility. You can't use it as though it was something that existed - it is just a statement - so watch out for it. Keep your eyes peeled and don't let anybody fool you with some fun-sounding story.

What about arguments that infinity can "sort of" be reached?

You can't sort of accomplish an impossibility.

THE UNIVERSE ISN'T JUST A BUNCH OF ROCKS

Fun-sounding word games regarding infinity

There are many "word games" that philosophers and others in the past have played. They involve the statement: "*If there was an infinite number of something...*" That implies that you have actually reached infinity. That's where the discussions become absurd and the reasoning statements are really just *word games* that we play.

For example:

If there were an infinite number of people, then there would be an exact duplicate of each of us. In fact, there would be an infinite number of exact duplicates of each of us. This means that there would be an infinite number of copies of you and an infinite number of copies of me and that is two infinities which is also against the definition.

Another example:

If an infinite number of time periods have existed, then every situation would have already happened, including this one, right now. Of course, it would have happened an infinity of times too. The logic in this instance is an illusion since you are assuming impossibility and logic doesn't apply - logic can't be used with impossibility.

Or - if an infinite number of time periods have always been, what is the number?

It just gets to be silly to think of it in any terms except imaginary, fantasy, or fun to consider after a few beers with the gang in college, or drinks after work at the bar.

INFINITY IS IMPOSSIBLE

Zeno's paradox using infinity

A famous Zeno's paradox involves Achilles and the tortoise and the two are going to run a race. Achilles, being confident of victory, gives the tortoise a head start. Zeno supposedly proves that Achilles can never overtake the tortoise. Basically, what Zeno is saying is:

Before Achilles can overtake the tortoise, he must first run to where the tortoise started. But then the tortoise has crawled to the next point. Now Achilles must run to this point. But the tortoise has gone to the next point, etc. Achilles is stuck in a situation in which he gets closer and closer to the tortoise, but never catches him.

One writer said this, concerning Zeno's paradox: "*What Zeno is doing here, is to divide Achilles' journey into an infinite number of pieces. This is certainly permissible, as any line segment can be divided into an infinite number of points or line segments. And, then he goes on....*" The problem with this reasoning is that the definition of infinity seems to indicate that you **can't** divide a line segment into an infinite number of anything since there isn't an infinite number of anything. There **is** an infinity of points - meaning that the number is endless. Lots of people get confused, including *an infinite number of mathematicians*.

Actually, you won't ever catch the tortoise if you try to do it this way. So, don't try. Have some turtle soup, instead.

THE UNIVERSE ISN'T JUST A BUNCH OF ROCKS

The famous *Halfway at a Time to the Wall* paradox

Aristotle stated a number of paradoxes such as the Dichotomy Paradox, in which he stated "That which is in locomotion must arrive at the halfway stage before it arrives at the goal."

The fact is that you **don't** have to go one-half of the distance, **each** time before reaching the goal.

Let's take a look at an example of this so-called paradox by stating it in this way:

"I want to move toward a wall which is ten-feet away. Before I can go the ten feet, I must go one-half of the way (five feet). Then, before I go the next five feet, I must go halfway again, or two-and-a-half feet. If you continue this reasoning, it can be argued that I must travel an infinite number of one-half distances before I get to the end of the ten-foot journey."

The common statement after posing the above problem is: "Therefore, I will never get to the wall."

The above problem assumes that we can move in one-half distances when we actually move in specific, measured distances. We move at a rate of, let's say, one-foot per second. Now, the time to go the ten feet is ten seconds. The time to travel all of the halves that you can dream of, and even more, is not infinity, it's ten seconds.

If I try to move toward the wall by moving "one-half-at-a-time moves," I will get a headache from the jerking. I will also have difficulty moving at the increasingly short distances. So, let's change to theoretical moving and calculate the "half-at-a-time moves."

INFINITY IS IMPOSSIBLE

It would appear that I am moving toward the wall, closer and closer and that, after an infinite number of moves, I will get to the wall. But, this is *wrong* since you *can't* move an infinite number of moves. Remember, you never get to infinity since there is always, at least, one more on the way to infinity.

So, I keep moving, each time for one-half of the remaining distance.

I do it for 26 million, billion, trillion, quadrillion, zillion, gazillion, lotsmorillion years.

And, I never get there. This is because, after I move one half of the remaining distance, I still have one half of that distance left.

I know, you are saying that by this time I'll have a long beard and, with a little breeze, my beard will hit the wall and this is probably true. But, my beard didn't get to the wall by doing "halfsies."

You will never go the ten feet by doing it one half of the remaining distance at a time. Never, never, never! So, you better be happy that you are moving by "wholesies."

Hilbert's Grand Hotel paradox

This is a fun story. However, as a serious story, it is a fraud and an insult to thinking, reasoning people - especially to mathematicians. (Hilbert was a very exceptional mathematician.) It starts with an erroneous, illogical statement and then uses mathematical jargon to try to convince people that it is valid (like many scams on the Internet do today.)

Paradoxes involving infinity aren't really that - they are just examples of what can happen when you start with an error and go on.

THE UNIVERSE ISN'T JUST A BUNCH OF ROCKS

The supposed paradox may be stated this way:

Consider a hypothetical hotel with a count of an infinite number of rooms, all of which are occupied - that is to say every room contains a guest. One might be tempted to think that the hotel would not be able to accommodate any newly arriving guests, as would be the case with a finite number of rooms.

Suppose a new guest arrives and wishes to be accommodated in the hotel. Because the hotel has infinitely many rooms, we can move the guest occupying room 1 to room 2, the guest occupying room 2 to room 3 and so on, and fit the newcomer into room 1. By repeating this procedure, it is possible to make room for any finite number of new guests.

It is also possible to accommodate an infinite number of new guests: just move the person occupying room 1 to room 2, the guest occupying room 2 to room 4, and in general room n to room 2n, and all the odd-numbered rooms will be free for the new guests.

There is more to this story but it all follows the same line of reasoning.

Analysis of the Grand Hotel paradox

Many analyses of this supposed paradox use all kinds of complicated mathematical analysis to show what is happening.

The story starts with "there are an infinite number of rooms" - which the definition says that you can't have. When you assume that there is an infinity of anything, the result will be meaningless and you can come up with many (an infinite number?) of results, none of which are really proven at all. It is all just a fairy tale of misleading words.

Aristotle's *Paradox of Place* paradox

INFINITY IS IMPOSSIBLE

Aristotle stated ".... if everything that exists has a place, place too will have a place, and so on ad infinitum." This seeming paradox was based on the use of "everything," which is another use of infinity and is, of course impossible.

And.........

Some theologians and science-fiction writers describe infinity as being boundless or without bounds. This sounds sexy but, again, is a play on words since infinity just isn't.

It might help to try this question on someone:

"If two plus two equals five then what does four plus four equal?" (Maybe, add for exactness, "Using the base 10".)

They may guess, as an answer, that four plus four would then equal ten or something else.

The correct answer to the question is:

"We don't know and we can't even answer the question since it went off of the diving board into absurdity after the start of the question where we said "If two plus two equals five ...."

Since two plus two doesn't equal five, the rest of the question is *non-recyclable garbage*.

THE UNIVERSE ISN'T JUST A BUNCH OF ROCKS

Georg Cantor's infinite sets

Georg Cantor was a German mathematician who came up with some *infinite set* theories in the late 19th century. His *proofs* are well respected by many mathematicians the world over. He tried to prove things about *infinite sets*. He tried to prove that the set of real numbers (any number) was more numerous than the set of natural numbers (any integer). His proofs were mostly based on theoretical *trends* of numbers. In other words, if numbers of one type continued to be more numerous than of another type, this would continue for all of infinity. His proofs are essentially accepted by many mathematicians.

What was meant by the term "infinite sets" was to include all of something which goes on forever - has no end. For example, all *even numbers*. Well, you can't include an infinite number of anything since infinity, as a reachable goal doesn't exist. To prove things about infinite sets, terms were invented like *uncountable numbers*. In infinite sets, of course you can't count things since they don't have numbers since the number doesn't exist.

His proofs simply must be invalid since there aren't infinite sets. They, like anything else infinite don't exist and how can you prove something in an impossible situation?

Also, I know one thing. Georg Cantor never said that you could reach infinity, just that you could "prove" things about infinite sets - which don't exist. (Sorry Georg. It really appears that you goofed, big time!)

Mathematicians' theoretical treatment of infinity

INFINITY IS IMPOSSIBLE

In mathematics, a symbol for infinity (∞) is sometimes used. This symbol cannot be used like other variables. Also, you cannot say that X divided by ∞ equals 0 or X divided by 0 equals ∞. The fact is that you cannot divide anything by 0. This symbol for infinity is just a placeholder for a definition, not a *thing* and should be used to warn you that this is a definition for something which is impossible.

Discussing dividing by zero is somewhat similar to discussing infinity. As you make the denominator smaller and smaller, the answer gets bigger BUT it doesn't reach or ever approach infinity.

To better understand, let's consider division in arithmetic. Let's say that we want to divide 1 by 2. Let's write it as ½. The numerator, 1, is a count of apples, for example. The denominator is a *rule*. What we are saying is that we want to divide 1 apple into absolutely equal parts such that there would be 2 of the parts. Dividing 3 apples by 2 would be saying that you wanted to divide 3 apples into equal parts so that there would 2 groups - of 1½ apple each. If you tried to divide a number of apples by 0, you would be saying that you wanted to divide the apples into equal parts so that there would be 0 (none) of them. This is, of course, nonsense and can't be done. I once heard a respected and competent mathematician say that dividing any number by 0 was *undefined*. That's at the least very misleading and at most just wrong! It is not *undefined;* it's nonsensical verbiage.

You just can't divide by zero! No matter how much you cry and pout and threaten, you can't divide anything by 0. You can't get away with saying "If you could divide something by 0, it would equal infinity" - because you can't.

You simply can't divide anything by 0.

THE UNIVERSE ISN'T JUST A BUNCH OF ROCKS

Most mathematicians are smart enough not to say "when you reach infinity" but to say, instead "as you approach infinity." Even this is not correct since you can't approach something which doesn't exist and is impossible.

The way that mathematicians usually say this is:

$1/n = 0$ as n approaches ∞. (Then, they say "the limit of," which means "as you approach.")

They assume that the *trend* will continue forever, which is what infinity is - forever.

In conclusion regarding infinity in mathematics:

Nobody, not anybody - at least nobody who has a considerable amount of brains - will argue that you can ever actually reach infinity. Of course, this is because you can't.

More on Aristotle and infinity

Many philosophers have contemplated infinity. For example, Aristotle decided that infinity had to exist because time appeared to have no beginning and no end. He came up with a great solution - that infinity both existed and didn't exist. He argued that, there was infinity and just potential infinity - infinity that could in principle be, but in practice never was. Again, time that doesn't stop, *in the future* is one thing but time that has always *existed* is not potential, it's against the definition of infinity.

Other examples of scientists *assuming* infinity:

Algebra example

A competent mathematician, who was lecturing on algebra, offered a proof that 3.9999.... equaled 4.

The logic used was:

INFINITY IS IMPOSSIBLE

X = 3.9999....

Then, 10X = 39.9999....

So subtracting, then

$$10X = 39.999....$$
$$\underline{X = 3.999....}$$
$$9X = 36$$

He was saying that .9999.... Subtracted from .999.... is equal to 0

This is incorrect, of course. You can't subtract infinity from infinity - you can't treat infinity as an *it*!

Electromagnetic force examples

Here are three examples of a common use of the word infinity to indicate that something doesn't ever stop. It is understandable as to what these people are trying to say, that it just goes on and on - but this is incorrect, and misleading terminology:

1. The strength of the electric force between two charged particles decreases as they get farther apart, according to the square of the distance. That is, if you double the distance between two charged particles, the force decreases by a factor of four. Because the force between two charged particles never drops all the way to zero, regardless of how far apart they are, we say that the range of the electromagnetic force is ***infinite***.

2. There are four fundamental forces at work in the Universe: the strong force, the weak force, the electromagnetic force, and the gravitational force. They work over different ranges and have different strengths. Gravity is the weakest but it has an ***infinite*** range. The electromagnetic force also has an ***infinite*** range.

3. Electromagnetic force and gravitational force both have an ***infinite*** range.

THE UNIVERSE ISN'T JUST A BUNCH OF ROCKS

A quote from a statement regarding the Heisenberg uncertainty principle, incorrectly using *infinity*:

"The reason this leads to a force with an *infinite range* is that given a limitless amount of time, a virtual photon can wander *infinitely far* from the charged particle it sprung up from, before being absorbed by some other charged particle. A virtual particle with mass will have only a finite lifetime (the heavier it is, the shorter the lifetime) and so it will only have a finite range. We'll see this with some of the other forces."

A statement of string theory's power of unification, example

Another quote, incorrectly using *infinity*:

"String theory, as stated above, postulates the existence of tiny vibrating strings that correspond to the observed elementary particles. Strings can undergo an *infinite number* of different vibration patterns, called resonances, whose evenly-spaced peaks and troughs fit exactly along its spatial extent."

The Universe and time periods have not always existed

The Universe consists of matter, space, time and Universe rules. Space isn't nothing; it takes time to go across a mile of it. Universe rules are the ones like *gravity, motion continues until something changes it, time slows with gravity, electrons fly around, atoms collect in groups as molecules, etc.*
If the Universe has always existed, this says that it has existed for an infinite number of time periods. In a real sense, infinity does not exist. Therefore the Universe, including time, has not always existed.

INFINITY IS IMPOSSIBLE

Let's say that you visualize a series of counts of hourly time periods, starting now, and the physical world, including the time-measuring system, is maintained. This count will then go on forever.

If you look backward at the past, and you assume that the Universe and time periods have always existed, there must have been an infinite number of time periods up to now.

You will then have actually reached infinity and you will remember, if you haven't been dozing, that you can't do that, period.

By the way, in theory you can talk about having an infinity of future time periods since this is simply stating that they go on forever, that they don't cease. However, you still will never reach the end - *for sure*.

The fact is that infinity of time periods as a real thing, that is reached - cannot be. There is strong evidence that this must also be true of all physical infinity. There are cases, such as "Black Hole" Theory of Relativity equations involving descriptions of finite mass with zero size. These do not necessarily mean that physical infinities exist; only that theory can't explain the situation - right now.

Time periods are of our Universe, involving matter.

Now, if time cannot have always existed for an infinite number of time periods, then it follows that the Universe must have started at some point. We can assume that it was the "Big Bang," many "Little Bangs," or some other way that started the Universe. However, there couldn't have been an infinite number of time periods.

The same reasoning will tell us that the amount of space in the universe is not unlimited. If it were, this is an infinite amount of space.

THE UNIVERSE ISN'T JUST A BUNCH OF ROCKS

In my book, "If I Had Been Born a Muslim" (2010), I used the argument that "Infinity doesn't exist and can't, therefore be reached" in order to <u>prove</u> that the Universe and Time periods have <u>not</u> always existed. I feel that this is a useful, valid and conclusive argument.

God as a supreme being can't do *everything*

The statement is often made that "God can do anything and everything." Normal reasoning tells us that he can't.

This is **not** that the Supreme Being is lacking in capability or isn't powerful enough, it is just that the word "everything" is another statement of an impossibility. It, like infinity, means that there is no limit and the number of things is *endless*. You can't treat "everything" as a "thing" or an "it" because it isn't either one. Remember, it is just a statement of impossibility.

What to watch out for about infinity

You can't reach infinity so don't let anybody tell you that you can, even for a moment or to get in good with your date of the evening.

Don't let anybody start a statement with "If there were an infinity of......" Just put your brandy glass and your foot down and shout "There ain't any! There ain't, there ain't!" (Maybe your use of the word forbidden by your old English teacher, Miss Sauerbraten, will shock them into consciousness.)

Don't let anybody use lots of fancy sounding mathematical terms to confuse you regarding infinity. Just tell them that "You'd prefer it if they didn't use that kind of language in your presence."

There isn't any such thing as paradoxical infinity or theoretical or conceptual infinity or possible infinity or axiomatic infinity or mathematical infinity or approaching infinity or twice infinity.

INFINITY IS IMPOSSIBLE

Voltaire said, "The only way to comprehend what mathematicians mean by infinity is to contemplate the extent of human stupidity."

Albert Einstein said, "Two things are infinite: the Universe and human stupidity; and I am not sure about the Universe."

Finally, don't let anybody use some other word which is also endless such as anything, everything, or forever.

But -

It does seem to be permissible or at least excusable, to say such things as "My love for you is *infinite*" or "I love you more than *anything*" or "I will go to *any* lengths for you."

THE UNIVERSE ISN'T JUST A BUNCH OF ROCKS

4 The Origin of the Universe

Biblical account

The biblical story of how God created the Universe in seven days is related in the book of Genesis.

According to the writers, God created the waters of the Earth, then the sky and finally the dry land. After also creating the sun and the moon, he worked on the task of creating plants, sea creatures and finally a man and woman.

In detail:

>**Day 1** - God created light and separated the light from the darkness, calling light "day" and darkness "night."
>
>**Day 2** - God created an expanse to separate the waters and called it "sky."
>
>**Day 3** - God created the dry ground and gathered the waters, calling the dry ground "land," and the gathered waters "seas." On day three, God also created vegetation (plants and trees).
>
>**Day 4** - God created light for the Earth (probably meaning the sun and moon) and to govern and separate the day and the night. These would also serve as signs to mark seasons and years.
>
>**Day 5** - God created every living creature of the seas and birds, blessing them to multiply.
>
>**Day 6** - God created the animals to fill the Earth. On day six, God also created man in his own image and woman. He blessed them and gave them every creature and the whole Earth to rule over, care for, and cultivate. He said that it was "good."
>
>**Day 7** - God having finished his work of creation, then rested on the seventh day, blessing it and making it holy.

This story reflects the attitudes and scientific knowledge of people many years ago.

THE UNIVERSE ISN'T JUST A BUNCH OF ROCKS

In modern times, people have learned many things about the Universe. They have plotted many galaxies including the Milky Way, our own galaxy, and our solar system including our sun and associated planets.

There has seemed to be solid proof of a long-term development process in the formation of the Universe and of our own Earth. This makes the bible story's exact timeline somewhat oversimplified.

The Big Bang Theory

In fairly recent years, much of the scientific world has adopted the concept that the origin of the Universe might have followed what is called the "Big Bang" concept.

The Big Bang model is the prevailing cosmological theory of the early development of the Universe. The theory purports to explain some of the earliest events in the Universe. According to the theory, the Universe was once in an extremely hot and dense state that expanded rapidly (a "Big Bang"). As there is little consensus among physicists about the origins of the Universe, the Big Bang theory explains only that such a rapid expansion caused the young Universe to cool and resulted in its present continuously expanding state. According to recent measurements, scientific evidence, and observations, the original state happened around 13.7-billion years ago. This can be referred to as the time that the Big Bang occurred.

Georges Lemaître first proposed what became known as the Big Bang theory of the origin of the Universe; he called it his "hypothesis of the primeval atom." The framework for the model relies on Albert Einstein's general relativity and on simplifying assumptions (such as homogeneity and isotropy of space). The governing equations had been formulated by Alexander Friedmann. In 1929, Edwin Hubble discovered that the distances to far away galaxies were generally proportional to their red shifts - an idea originally suggested by Lemaître in 1927. Hubble's

THE ORIGIN OF THE UNIVERSE

observation was taken to indicate that all very distant galaxies and clusters have an apparent velocity directly away from our vantage point: the farther away, the higher the apparent velocity.

If the distance between galaxy clusters is increasing today, everything must have been closer together in the past. This idea has been considered in detail back in time to extreme densities and temperatures, and large particle accelerators have been built to experiment on and test such conditions, resulting in significant confirmation of this theory. On the other hand, these accelerators have limited capabilities to probe into such high energy regimes. There is little evidence regarding the absolute earliest instant of the expansion. Thus, the Big Bang theory cannot and does not provide any explanation for such an initial condition; rather, it describes and explains the general evolution of the Universe going forward from that point on. The observed abundances of the light elements throughout the cosmos closely match the calculated predictions for the formation of these elements from nuclear processes in the rapidly expanding and cooling first minutes of the Universe, as logically and quantitatively detailed according to Big Bang nucleosynthesis.

[Nucleosynthesis is the process of creating new atomic nuclei from pre-existing nucleons (protons and neutrons). It is thought that the primordial nucleons themselves were formed from the quark - gluon plasma from the Big Bang as it cooled below two trillion degrees. A few minutes afterward, starting with only protons and neutrons, nuclei up to lithium and beryllium (both with mass number 7) were formed, but only in relatively small amounts. Some boron may have been formed at this time, but the process stopped before significant amounts of heavier elements such as carbon could be formed.]

Fred Hoyle is credited with coining the term Big Bang during a 1949 radio broadcast. It is popularly reported that Hoyle, who favored an alternative "steady state" cosmological model, intended this to be pejorative, but Hoyle explicitly denied this and said it was just a striking image meant to highlight the difference between the two models. Hoyle later helped considerably in the effort to understand stellar nucleosynthesis. After the discovery of the cosmic microwave background radiation in 1964, and especially when its spectrum (i.e., the amount of radiation measured at each wavelength) was found to match that of thermal radiation from a black body, most scientists were fairly convinced by the evidence that some version of the Big Bang scenario must have occurred.

The Universe has always existed

Some argue that the Universe has always existed. I must assume that this theory doesn't have to wrestle with the question of how things got started in such a profound and orderly manner since "It just always was."

We have already clearly stated that *infinity* is an impossibility and, therefore, the Universe couldn't have always existed. (If it had always existed, there would have been an *infinite* number of time periods.)

What matters the most?

It is very important to note that these concepts don't really relate to the question of the origin of the Universe, but only to some of the steps in the process.

Theories of the origin of the Universe usually take the simplistic viewpoint that the most important issue in explaining this is in the explanation of how all of the matter and energy managed to get created, expanded and distributed throughout the vast regions of space.

THE ORIGIN OF THE UNIVERSE

This is "grade school science." While this is a very interesting subject, it very definitely isn't the big, important question. Spending years of effort in trying to establish the validity of the Big Bang theory is interesting, but begs another, most important question:

What about the rules?

How did the rules (the laws) of the Universe get created?

It is also important as to how people were created, but I believe the basic, key question, to start with, is how the different *types* of things, which make up the Universe, were started. Then, the *real* key is how did the rules (or laws) get created or started?

Now, it is apparent that the rules involve matter in the Universe. So, it seems obvious that the rules couldn't have been created or started before matter/space and time started. Likewise, physical matter couldn't have started before the rules started. They all are tied together and fit one, coherent, integrated system.

In a like manner other elements of the Universe, like energy, light and magnetic waves and others, all work together as a package.

It is logically reasonable to assume that all elements of the Universe started (were created) at the same time. It just wouldn't make a coherent package to consider any as existing apart from the rest.

Was there a master planner/creator?

As previously stated, since the Universe didn't always exist, it makes sense that it didn't just start by itself - that there must have been a creative force that started it.

Now, you can try to use "chance and natural selection" if you want to, but it seems obvious that most reasoning people would throw that out in a New York minute. It is especially difficult to reason that the rules/laws were created through chance and natural selection.

Again, let's call this creative force, God. (Note: I believe that most religious beliefs usually follow this line of reasoning, although many do so with a creative, involved and expanded description of God.)

This creative force existed before the Universe existed and therefore wasn't part of the Universe. So, **God** isn't classed as part of the Universe. (It is reasonable to class God as both in the Universe and in the non-universe.)

Are there multiple universes?

In scientific literature, these days, there seems to be a thin line between science fiction and real science on this subject. A number of theoretical physicists and others have come up with complicated theories arguing that there must be a number of Universes. Some even argue that there are parallel universes where there are duplicates of each of us with some pretty far out complications. Some scientists try to use mathematical proofs even showing just how many other Universes there must be.

One question is: What did a creator have in mind when multiple Universes were created? Wasn't one difficult enough?

Another question is: What happened to solid science when scientists are proposing these ideas?

I can't say that these ideas are impossible but I can question them.

5 The Non-Universe

The Universe didn't always exist because then it would have existed for an infinite number of time periods.

If the Universe started at some point, then before this the *non-universe* must have existed.

About all that we know about this is that there was no time, no matter, no energy, no rules/laws, no mother-in-laws and no blondes.

Let's also assume that there must have been a planner/creator which I will call God.

When you say that there was no matter, this means more than just empty space. There wasn't any space either.

What it means is that there was a whole different *deal*! There was a completely different set of rules and laws. They wouldn't even be called laws or rules.

It is difficult to imagine, so really think about it. Change your whole way of thinking from the usual patterns.

Nothing is the same. Trying to place yourself in this frame of reference, you wouldn't necessarily even *think* the same. You probably wouldn't think logically or iteratively (although you *may*). Remember, there is no *time*. There is nothing to think about, in the same way as we think now.

But, you must know that there are a whole new set of circumstances and rules, or the equivalence of the rules. It is almost impossible to visualize - in fact, it is absolutely impossible visualizing since there is no sight, of course.

THE UNIVERSE ISN'T JUST A BUNCH OF ROCKS

One other thing, you, as God, can probably understand the Universe too, even though it is not yet created, since it will be - by God.

That's about all that we know about this situation. It isn't surprising that we don't know much about this situation since there wasn't anything which we are familiar with to know about.

How can you know very much about nothing?

However, you must suspect that there was more to it than just nothing, we just don't know much, but that has already been stated.

Well, do you get the picture about the non-universe? (Of course, there was no *picture*.)

Oh, and a couple of other things. First, assume that God is able to exist in this non-universe. And, since there is no matter, it seems logical to assume one other thing: "If there is a heaven, where souls go after death, this is probably where it is. Right here in the non-universe! Wow!

Let us theorize and guess about the non-universe

What we know about the non-universe is really limited.

What about the happenings in this non-universe? What about the origin of this non-universe and of the creator?

Let's just make some wild guesses about this.

I have said that there isn't any time in this non-universe, but that only applies to our matter/space/time Universe type of time periods. Another definition of a type of time is the order of happenings. It is reasonable to theorize that whatever the processes occurring that there may have been happenings that had some *order* to them.

THE NON-UNIVERSE

So, maybe the process of creating the Universe had some happenings where the order was important.

If this is true then this "order of happenings" type of time must have had some start also, or there would have been an infinite number of them - which is against the definition.

The easiest way to think about this process starting is to think that the happenings, with an order, just *started* at some point. Perhaps they started because it was important and, before that, there was no order. In fact, there could have been several periods where the order of happenings was important and so, these special type of time periods started and, perhaps ended.

It is convenient to think about the creative process (God) as operating like a *brain* or, perhaps with some kind of *intelligence*.

Whatever kind of theories used concerning this *brain* or reasoning process, I must concede that it is far different from our human *brain* (or some computer-type of reasoning ability).

It certainly wouldn't seem reasonable to assume *words* would be used by it, for example. Nor any other human-type thought processes such as memory recall, either.

The brain or intelligence idea might be OK to use as a general concept but there is no reason to assume that this would be anything like our human ones.

The whole set of rules and occurrences in this non-universe would probably be completely different from our human existence here on Earth.

An interesting thought is this:

THE UNIVERSE ISN'T JUST A BUNCH OF ROCKS

There has been much discussion as to whether humans have free will or whether determinism is solely responsible for all actions and occurrences in the Universe. It would seem that there can be no question at all that the creator of the Universe (God) must have free will. Isn't that really obvious?

What about the start of this non-universe?

I do not think we need to be concerned about the start of this non-universe since there wasn't any beginning since there wasn't any time. The time periods of our Universe didn't exist and the type of time considering the "order of happenings" is assumed to have started and stopped, perhaps, many times.

What about the reason for its existence?

It seems reasonable and, certainly desirable to assume that something must have a purpose or a reason for its existence. However, logical reasoning doesn't confirm that this is necessary

What about the idea (a wild guess) that the non-universe simply had to exist because otherwise there would have been absolutely nothing and that just wasn't, sort of, allowed?

Also, why was the Universe created? An often quoted reason has to do with the Universe and people being created to test their ability to perform right or wrong or good or bad. Many prophets and authors have propounded many ideas - often supposedly sent from God. However, God, himself has never sent a video to explain exactly what the deal is. Maybe, leaving things somewhat open is actually part of the deal.

There are many fantastic ideas that I could come up with but I must admit that I *just don't know* and that's it - in spades.

THE NON-UNIVERSE

*In summary, the one thing that I can be fairly confident about is that the processes, the description, in fact, **everything** in the non-universe would seem to operate differently from our Earthly existence.*

It would seem that attempts to make the non-universe or describe God as operating in any similar ways to our human existences would be ridiculous. Only our human egos would allow us to even think about such a thing.

Perhaps that is why we have never seen an appearance of God, himself, here on Earth.

THE UNIVERSE ISN'T JUST A BUNCH OF ROCKS

6 Can You Describe the Creator of the Universe?

Let's step back and take a look at what we know about God. I must assume that there was a planner and a creative force which started the Universe, which I will call God.

I believe, through my personal experience, that there is an "answerer of prayers."

In addition, there are, of course, many accounts by many representatives of the many religions, of God's existence. There are many descriptions in the Bible, in the Koran and in many writings by various people of an understanding of a God.

God as the creator of the Universe

Most people will agree, and documents attesting to people's religious beliefs confirm, that no one has ever seen God in person.

However, **we do know quite a bit about God, the creator**.

First, let's assume that this being doesn't have a natural physical form since he existed before the universe and there was no matter then. It is no wonder that we haven't seen him, although there are many accounts of different religions including "angels" or some representative showing up as "men, and in physical form."

I am referring to the situation before God created the Universe, the non-universe. In this situation, there is no time or energy or matter or any of the elements of the Universe. Remember, **NO TIME!**

God, in this situation, existed with a whole set of new conditions other than the *Universe conditions*.

THE UNIVERSE ISN'T JUST A BUNCH OF ROCKS

Then, God created the Universe and must have been able to **also** coexist and deal with the Universe conditions, including time, energy, matter and all of the rules associated with this Universe.

Do you begin to get the feeling that God is pretty special in capabilities?

Awesome is more like it, I would say.

I am making the assumption that this "being" is a lot more capable than "Joe, the bartender" or the movie star of the week or, even you and I. He did start the whole thing of the Universe including the rules that tie everything together and can handle both the non-universe and the Universe - WOW!

Let's just think about the planning of the universe. He set out the entire myriad of rules and made all of the decisions of equations that describe time, space, matter, velocity, gravity, molecular activity - and on and on. Then, he put it all into motion with the "big bang" starting matter expanding at almost the speed of light (really faster than this speed considering that space was expanding along with matter).

"God" then watches the unfolding of elements in the Universe over billions of years.

It is apparent that "God" chose evolutionary development over time as a tool.

So, he watches planets evolve, organics progress and, finally, people evolve. Then, people progress (I guess that I can use the term, *progress*). There are wars and struggles. There are famines, earthquakes and many disasters. There are technological advancements. Populations explode. Religions evolve using good and bad techniques and do both good and bad things over years and years and years.

CAN YOU DESCRIBE THE CREATOR OF THE UNIVERSE?

Now, one thing of note: It isn't clear if "God" is totally "tied" to **time**. Remember, he existed before time was started. So he may be watching as these things evolve and people progress (again, the word "progress" is used loosely). He may also be observing, as only "God" can without being tied to our time. He may be outside of our "time" system and observing in another way.

A reasonable theory is that he is capable of both observing in our time frame and, also, outside of it, in a "non-time" mode. Again, he can exist and operate both in the non-universe and in the Universe mode.

I am considering that "God" is pretty darn capable. Just to start with, think of all of the computers in the world today. It is reasonable to assume that "God" might have, as a starting point, the capabilities of, not only computers today, but also all of those of hundreds of years from now. These will be capable of doing things like performing virtual reality, doing all of the things that the Internet will do, multiplied by many, many times over - hundreds of years from now - and on and on.

A fantasy story about the creation of the Universe

An idea is presented here which is presented just to make you think.

God possessed fantastic capabilities. This whole concept is of a being who was/is way beyond any of our human capabilities.

It is difficult to imagine such a being. To start with, he must be in a whole different realm than we are.

So, just to dream about possibilities, let's think of a picture:

In this picture, God is a child in a really, really advanced family.

THE UNIVERSE ISN'T JUST A BUNCH OF ROCKS

He is in the non-universe with no physical things like in our world but a whole set of different things with a whole different set of rules. Everything is different, it being a non-universe, of course.

So, it's a *Sunday* afternoon (*sort of*). God, the "child" is bored and looking for something to do.

He thinks: "I guess that I'll make a Universe". He says, "I guess that I will set up things using some unified concepts. All of the rules will fit together and interact in a way so that everything works together, seamlessly. I think that I'll invent Time. This will make things change with this Time. Then, let's have Space and Gravity and everything will fit together. I'm going to invent all kinds of little tiny parts to fit together to make bigger physical things. (Like molecules, atoms, electrons, quarks, energy strings....). I'm going to develop objects and regenerative/reproduction so that the process will really expand without having to redo every little thing over and over.

"I have a plan for what I'm trying to do so this will really work out. I'm going to try to prove *good* and *bad* and, I think, I'll make some human beings to see if they will do these things. I'll even reward them if they do right - yes, this will be my plan!"

He says, "Hey, this is fun."

His *Mother Equivalent* calls him and tells him that he has an *appointment* in a short while.

He thinks, "I guess that I'll use *evolution* as a method to make things progress in this Time while I'm off busy with my other things. This will make things develop sort of by accident, with things changing and developing according to how things are working out (survival of the fittest).

"But", he thinks, "I'm going to keep inputting my guides for things, too, so they work out as I want in my overall plan."

CAN YOU DESCRIBE THE CREATOR OF THE UNIVERSE?

He sets up things to start and run by themselves, somewhat but keeps coming back to nudge and change things so they fit his overall plan.

Conclusion of the fantasy story

The preceding is quite fantastic and, considering in a reasonable manner, it, of course, is not right on target.

However, some, equally outlandish idea must be behind the creation of the Universe by the Master Planner and Creator.

Now, imagine yourself as God for a moment, if you can.

This will take considerable imagination but try to get into it.

You are God in the non-universe. You see things without time being involved. You have no physical form, of course, since there isn't anything physical.

Then you create the Universe.

Then, you are also God in the Universe. Now you see the Universe evolving as it has for some 14-billion years. You see things in a flash that take place over these billions of years.

You see wars, people clashing and evolving. You see good and bad.

You see it all, and understand things.

You also see how things fit into your plans - you must have had some reason for doing all of these things. You must have had some plans. For example, to see what would happen because you allowed *free will*. Perhaps other plans too advanced for us to understand, or - let your imagination run wild!

THE UNIVERSE ISN'T JUST A BUNCH OF ROCKS

You see all of this and you *understand it*, or you wouldn't have created it in the first place.

You are not awed by yourself since you know and understand why you are here (probably). You know that you have the abilities to do the things that you do.

You, as God, probably, also have some guiding thoughts about your actions. You want to remember to check and recheck your plans.

You keep examining what you have done and what has happened to see if you should adjust any of your plans or actions. What about this free-will thing? What about the suffering that some people are going through? What about the exceedingly long time those things are taking? (Does time matter to God, probably not, after consideration.)

Also, what about that *one person* who is now reading this book or the *author* and that person is thinking about these things? Should you be giving that person some special consideration? Should you consider letting that one person have special knowledge so that they could do special things to help your plan along. Should they have special knowledge that would let them move mountains or be able to influence powerful people? Should they be able to solve the problems of exceeding the speed of light or be able to solve something else?

There are lots of possibilities, especially since God (you for the moment) has so many immense capabilities.

OK, you've got a start at imagining what you would feel and know as God, so think about it for a moment.

Now, let's ask you/God some questions.

CAN YOU DESCRIBE THE CREATOR OF THE UNIVERSE?

Are you going to be absolutely enthralled with the idea that people should worship you? Do you even begin to need that? (Note: In the past, I have felt really thankful and pleased when God has answered some of my entreaties, but, I'm not sure that God would need or even want my praise and thankfulness.)

Do you want people to sacrifice lambs, or something else, on the altar?

Are you going to stop people from building a tower of Babel by making them speak different languages? Are you going to get mad at the world and destroy it in Noah's flood? Couldn't you have solved the problem in a more advanced way?

I only ask these questions to show how some of the old biblical stories sound more like people's simplistic view of what God might be rather than what you feel like when you are "sitting in God's fantastically awesome position" for a moment.

Now, another question:

You want to send Jesus to earth to teach people how to get along. This is important. Do you need to have a virgin give birth using, I assume, your special DNA? What DNA? There isn't any matter or energy in your realm. Do you need to ask the virgin's permission before performing this act? Do you need to announce the birth of Jesus by sending angels down to tell shepherds about it? Will you send *Wise Men* to attend Jesus, shortly after birth? These things are very good at representing what people might imagine that God might do, but would the God that I have described do things like this?

The questions go on and on.

How do you, sitting as God, for the moment, feel about the questions?

What is your position? What makes sense?

THE UNIVERSE ISN'T JUST A BUNCH OF ROCKS

Put your thoughts into words and write them down on a piece of paper.

What *can't* God do?

It is clear that God, the creator, is fantastically capable and understanding.

Many people jump from this understanding to the position that God can do **anything**.

Well, he can't do anything and everything.

The position that he can is not just wrong, it is logically unsound and absurd.

This is another use of a definition just like i*nfinity can't be reached*.

The definition of anything and everything is that, no matter how much is done, there will always be more - so you can't reach the end.

The same logic which has been used to show that infinity can't be reached is applicable here.

For example, if God could do anything, he could set up a rule that gravity will **always** work and be in place in any situation. Then, he could set up a rule that dictated that gravity would **never** be applicable.

These are contradictions.

It isn't that God isn't powerful or capable enough to do something. It's just that contradictions are against the definition of things and aren't logically allowed when you are using reason.

CAN YOU DESCRIBE THE CREATOR OF THE UNIVERSE?

Also, if God could do absolutely anything and everything, he could carry out whatever his plans for creating the world were and do this instantly without even having to create the world.

Think about that when you want to claim that God is absolutely omnipotent.

If he were, he could do everything RIGHT NOW and there would be nothing left to do.

This is just another example of how ridiculous it is to ascribe absolute omnipotence to God.

So, God can't do absolutely everything. This shouldn't bother you. If it does, you might want to consider getting help from a good psychiatrist.

(As another thought, I don't have enough knowledge about God's background to be sure that he doesn't have a "Master" or "God" that dictates things to him. The point, here, is that while I know some things about God due to the fact that he created the Universe, I don't know it *all* by a long shot.)

God's interaction with people

I've tried very hard to imagine what it would be like to see things from God's viewpoint. Now, what about any interaction with people?

If people are really important in God's view and more special than other animals and lower forms of life, perhaps God has set up functions in the Universe to handle interactions with people. (Following this line of reasoning, he could have also set up functions to handle interactions with other animals too.)

THE UNIVERSE ISN'T JUST A BUNCH OF ROCKS

For example, when we pray to God asking for help in some area, there could be "prayer objects" that sometimes answer our prayers. There might even be rules in the Universe such that our prayers get answered based on worthiness or need or whatever.

Whatever the answer, I, through personal experience have had my prayers answered many times in ways that went beyond the normal probabilities of "chance" in their help in my situations.

I don't have the answer to this but feel that there is personal, person- relationship with God that somehow works and shows results.

Will God answer all of people's requests (prayers)?

Why doesn't God **grant** all of people's prayer requests?

The answer is that there would be **chaos** if all prayers were granted.

Of course, one result would be contradictions. Sam might ask God to make Shirley smart and Bill might pray to have her made dumb. What's God to do?

If everybody got what they wanted:

No one would need to work, we would need no inventions or discoveries since one little prayer would do it. Everyone would live forever. We would all just sit on our front porch swings drinking margaritas.

Everyone playing the slot machines would win but the casino owners would pray for them to lose.

It just wouldn't work and that's the end of that!

CAN YOU DESCRIBE THE CREATOR OF THE UNIVERSE?

The real question is "Why does God answer *any* of our prayers?"

Now, in my personal experience, I have found that God **has** answered many of my requests. Something happens after I have prayed for it which *could* have happened without the prayer, but the probabilities are tremendously against it. I am convinced that these occurrences are because of some "outside" help.

A comment: Since I have asked for help with this book on several occasions, I feel that God has helped me with it quite a bit. An interesting thought is this: If you want to argue with my ideas, you might consider that you are arguing against God. Better watch that!

Now, the preceding was just for fun. If you want to really nail me, you could just say that God helped you when you were pointing out any discrepancies and that your complaints were against my errors created when I did things without God's input! There, you got me.

Of course, I can't answer the question as to why some of our prayers are answered, for certain. I can only make some educated guesses.

It may have to do with some "*rules for answering prayers*" that are set up. For example, how worthy you are, how fervent you are, how important it is to you - perhaps how it fits into the overall Universe plans, etc. (And, if you want to get a little carried away, maybe if you had some help from somebody who predeceased you.)

But, a word of caution:

God seemed to be "in love with rules" when he created the Universe. There is gravity, time, orderly electrons and lots of other orderly little and big things. There are rules like "If you step in front of a speeding car, you will get hurt." It goes on and on.

THE UNIVERSE ISN'T JUST A BUNCH OF ROCKS

It seems that a *Rule Oriented God* wouldn't want to violate his rules just willy-nilly.

So, be careful of asking for too many "miracles" that violate the overall plans. My feeling is that you won't get a positive response in **most** cases.

But, you already know that, from past experiences, don't you?

7 Universe Philosophy

Most of this discussion, to this point, has been based on logic and reason as the main emphasis.

Sometimes, it's important to consider some philosophical approaches in coming to conclusions also.

One of the methods that I have personally followed, with some success in the past, is what I call the *Phydefiteration* method. This, basically, consists of just the normal scientific process which is used by many people. I named it that because I have been unable to find it described in well defined, very specific terms, in the literature.

The Phydefiteration method

This method starts with a philosophical statement and uses an iterative process until a more specific set of philosophies or rules is found.

After successive iteration, the statement becomes more and more applicable until it is something that the user can really rely on, and use as a *general* rule.

For example, let's say that we had thought that we should really back up our computer work in computer storage. (That's our beginning statement.) Then, one day, our computer crashed, our disk was lost and we lost a lot of work therein.

So, we said, "Boy, I really should back up my computer, once in a while."

Then, it happened again and we lost two weeks' work. (We're really disgusted, now!)

THE UNIVERSE ISN'T JUST A BUNCH OF ROCKS

Our new statement is, "I should back up my computer every day." So, we did back up the computer, almost every day, but once it was forgotten for several days and the computer crashed again.

Now, we said, "I'll never again work on a computer - ever - without having an automatic backup system installed that will back up things every single day, come rain or shine."

This is the "rule" obtained through Phydefiteration.

Another example might be where it was theorized that God might answer a prayer. It worked.

The Phydefiteration philosophy is, "Whenever I have a really serious problem or need, I will ask for help in a prayer."

Another is, "If an e-mail comes with a story that includes statements like '*This is really true*' in it, it might be a scam. Later, this becomes, "It is almost always a scam."

Or, another example (one of my favorites) is, "In developing computer/user interfaces, the user is fairly important."

The Phydefiteration philosophy is, later, "In computer/user interfaces, we should go to the death in making sure that the user is absolutely the most important thing in the entire world, bar none."

Then, we can rely on well-developed Phydefiteration philosophies without even thinking, after they are developed and reinforced.

However, remember that this shouldn't be used as just b**lind faith**, without constant reconsideration and iteration, in the future.

Blind faith is the enemy of scientific reasoning

It is important that we all have teachers throughout life. Our most important teachers are usually, our parents and those whom we respect in our early years. Religious beliefs are often ingrained in us. Sometimes we are told that to question our taught, religious beliefs is heretical - and worse. Sometimes, it can be difficult to think above such things and look for the real truth, even in small insignificant details. The same thing is true with prejudices, of course.

Scientific reasoning sometimes points us in different directions from our early religious education.

In my previous book, "If I Had Been Born a Muslim," I tried to point out that our religious beliefs are quite different if we are born, for example, as a Muslim, a Catholic, or a Jew. I suggested that we should try, as best we can, to start over with most beliefs, since only when beliefs are really reasoned out are they real beliefs and we are not just being robots and repeating what someone said.

All that we can do is to try, since:

Blind Faith is the "Black Hole of reason corruption."

Blind Faith is a "jelly sandwich" which has no benefit to the healthy body in any way except that it tastes good. (I apologize to jelly home-canners or manufacturers.)

In case you have missed the point, I am trying to say that to thinking, reasoning people, *Blind Faith is* **rotten to the core**!

The place of religion in the Universe

Religion has always been part of society. Almost every civilization or society seems to have always had religion. In recent years, communist countries have tried to eradicate it, but people held to the old ways and worshipped in secret.

When was the first religion recorded? Recorded history only goes back 6,000 years or so. The earliest civilizations in Mesopotamia and the Indus and Nile valleys all had formal worship of some sort of supernatural critters as integral parts of their cultures. These, no doubt, had much earlier roots in pre-historic cultures. Religion and culture probably grew up together, one reinforcing the other. Religion probably got financial and political support from the organized society and the society used religion to control and motivate the population. In Colonial America towns grew up around churches. Something very like that probably happened in prehistoric times as groups with common beliefs banded together for mutual support and created the first communities and cities.

While much use of religion was to control people, there seems to have been something like an innate need or desire or belief for people to believe in a higher being.

This cannot be disregarded when trying to analyze the Universe, its origins and even its purpose.

The place for "feel" in reasoning

Remember, "feel" or "sense" is also important in reasoning.

If something doesn't "smell" right, be careful. You shouldn't count on the sense or the feel as proof of something; but it can certainly steer you in examining the directions that you take.

For example, it "feels" right that, throughout history, evolution had a "guiding hand" in steering the course through many developments of life processes. An examination of probabilities along the way may help to offer hints of proof that this wasn't totally a random process - it might have been against normally calculated probabilities for it to have been.

THE UNIVERSE ISN'T JUST A BUNCH OF ROCKS

8 The Object-Oriented Universe

What do we mean by "object-oriented?"

The term "object-oriented" is borrowed from computer programming.

In the early days of computer programming, programs were written with many "modules" or "subroutines" placed in "libraries." These subroutines were developed to perform commonly needed tasks so that the same programs didn't need to be written each time the function was needed. The use of the term "object-oriented" for modules started to be commonly used in the early 1990s. These "objects" were self-contained and performed all of the tasks in their defined nature when referenced by another program. A basic definition of object-oriented computer programs is as follows:

An object-oriented program may be viewed as a collection of interacting objects, as opposed to the conventional model, in which a program is seen as a list of tasks (subroutines) to perform. In object-oriented programming, each object is capable of receiving messages, processing data, and sending messages to other objects and can be viewed as an independent 'machine' with a distinct role or responsibility.

An object-oriented Universe

The Universe is, of course, different from a computer program. However, many of the same principles can be seen in the design of our Universe.

Let's start with quarks, an elementary particle (or "object"). These are used to make up protons and neutrons which, in turn, make up atoms - then elements, into molecules, etc. And, go all the way to galaxies (also "objects") which are held together with rules such as gravitational attraction.

THE UNIVERSE ISN'T JUST A BUNCH OF ROCKS

It seems that "God" as the causative force in the design of the Universe really was attracted to the idea of objects and to the use of rules to govern objects. All of these objects function independently while following the rules of the Universe. They then give us all of the many millions of different objects which make up our existence.

Look at it another way. If God had to build all of the things in our world and constantly tell them how to behave, he would be really, really busy. It appears that his design was a pretty smart one. In fact, it is so good that I would give him the architectural design award of the millennium - at the very least.

Now, I'm going to say something that may cause a lot of dissatisfaction among fundamentalists.

Please don't start to stone me until you, at least, hear me out.

Okay, here we go!

God didn't create you and me. Instead he set up the system so that our mothers and fathers could do it - while, I must add, they were having a really good time.

He also didn't create the fishes of the sea or the lions or the tigers or the cute little panda bears. He just set up the system so that their parents could do it - again, while they were whooping it up. This "whooping it up" is what keeps the process going too. Again, God didn't forget a thing.

The use of self-perpetuating objects doesn't end with reproduction. Oxygen is taken in by animals. We use it to allow us to walk to the bar and share a beer, martini, or mai tai with our buddies while discussing the game. Then it is expelled as carbon dioxide. Plants turn this into oxygen again, through photosynthesis, thank goodness, or we would all be deader that door nails by now.

THE OBJECT ORIENTED UNIVERSE

There are countless types and flavors of this use of objects in our world.

An interesting thing is that the Universe seems to be a little less complex and unimaginable when viewed as many, self-governing objects. The development of these objects, the grouping of these objects, and the rules around them is still not a trivial task. However, the whole development does seem a little more manageable with God's object-oriented approach.

What about God's interaction with people - prayer?

Let's start by examining what God would consider when thinking about interfacing with people.

Put yourself in his place, in the non-universe. You are sitting there (not really sitting, since there aren't any chairs, not having any matter). You have come up with all of those great, prize-winning ways to build the Universe by using objects and rules so that you don't have to handle every little thing, each and every time.

Now what? What about prayer? Are you going to have to listen to all of Adam and Eve's entreaties yourself? Then, being pretty darn smart, you realize that there may, someday, be thousands and maybe billions of Adams and Eves. Okay, I got a little carried away with symbolism. Let's start again.

God created the Universe using self-handling and perpetuating objects. It makes sense that the interaction with people through prayer would be handled in the same way.

As I mentioned previously, I believe in prayer and have had my prayers answered many times in a positive, beyond probability of it being an accident, way.

THE UNIVERSE ISN'T JUST A BUNCH OF ROCKS

Using reasoning, it may be that our prayers are handled on an "object" basis. Perhaps, many of the prayers are handled by some of God's rules rather than by a personal interaction with a supreme being on a one-to-one basis. Wouldn't it be interesting if our prayers were handled according to how serious we were about the subject or how serious the problem was that we were having; or maybe, even how ardently we prayed - or perhaps, how good we had been.

Whatever the answer, I believe that prayers do get answered. I, also have a *feeling* that there are some of God's rules involved in getting them answered. I'm really convinced, due to experience, that the importance and the justification of the subject have some bearing on how well they are answered. Finally, I have a strong feeling that justice is often involved. In other words, if you ask God to give your enemy a "hot foot," your prayer may very well *not* get answered in a positive way.

9 Can Evolution Totally Explain the Universe Make Up?

Evolution of life forms, including animals on Earth, is proven to have played a role in the development of life on Earth and, even of humans.

It is difficult to argue that there is **no** evolution of life forms, even of humans. There are ample fossil remains to clearly show that evolutionary development took place. It seems clear that this was the master planner's (God's) plan.

Accepting the obvious fact of evolution does not require that you accept that evolution is totally responsible for development, however. That is, that everything developed from *nothing* using only chance and survival of the fittest.

It seems somewhat questionable to explain **all** of the development of humans to this kind of evolution. This is because of the extremely complex and involved nature of humans and also because of the ascribed value to some theories concerning the human's relationship with God.

Evolution of the Universe, itself - I don't think so!

Considering evolution as a way to explain the makeup of the **Universe** is another question entirely.

It is difficult to even conjure up theories about how the relationship of space, matter, time and the myriad of "Universe rules" could have come about with evolution; through "**trial and error.**" Boy, how about the length of time **this** would have taken? You talk about billions of years - that's nothing compared to what would have been required!

THE UNIVERSE ISN'T JUST A BUNCH OF ROCKS

Based upon scientific "reverse extrapolation," it is estimated that the Universe has been in existence for about 14-billion years. Again, that's a long time but not when considering the task of evolving something like the Universe with all of the rules/laws and different elements. A few billion years seems far too short to randomly evolve the **rules** of the Universe!

Not only would it have taken a really long time but the whole idea of chance and survival of the fittest simply doesn't seem to fit with this concept.

Of course, time isn't really a factor since the Universe started (and it did, remember - *no infinity*). So, before it started, there was no time to worry about.

Take, for example, the rule (law) that "an object in motion will remain in motion until something alters it." Try evolving to that through trial and error or survival of the fittest.

The tremendous complexity of it all has driven many scientists to search - in vain - for a "Unified Theory" of the universe rules, so as to simplify the development problem.

Evolving gravity and space and time also seem *just a little difficult* to comprehend.

It seems that there must have been another explanation. A very reasonable theory is that there was some planning involved and this means there had to be a **planner**. This **planner** doesn't have to be exactly similar to a human being. In fact, a casual analysis of the necessary capabilities of a planner and creative force of anything like this magnitude indicates that this planner must be just a **little bit more capable** than any human that you or I have ever met or dreamed about.

CAN EVOLUTION TOTALLY EXPLAIN THE UNIVERSE MAKE UP?

As previously mentioned, since there was no matter in the non-universe before the Universe was created, this planner/creator wouldn't look like any human, of course. He wouldn't *look* like anything since there was no matter. But, don't worry about that. There were no eyes to see with, either.

*Note: The fact that there was no matter before the Universe started does **matter**, however.*

Evolution of humans

Considerable evidence of evolutionary development of humans and their ancestors has been discovered through fossil examinations.

It has been perhaps 2.5-million years since the appearance of the genus Homo.

Then, maybe 200,000 years ago, humans started looking like they do today and 25,000 years since Neanderthals died out.

It is believed that humans originated about 200,000 years ago in the Middle Paleolithic period in southern Africa. Seventy-thousand years ago, humans migrated out of Africa and began colonizing the entire planet. They spread to Eurasia and Oceania, perhaps 40,000 years ago, and reached the Americas 14,500 years ago.

Future evolution of humans

It is interesting to consider the possible, future evolution of humans.

Evolution, as it is usually considered, takes perhaps, at least one thousand years before any significant changes have shown up in people. Look at people 400 years ago. There are many differences in behavior and attitudes but not in physical, nature- ordered attributes. George Washington may have had wooden teeth but these were dictated not from his DNA, but from the lack of a dentist on the corner with a lab with good materials.

So, it is predicted that, at least for the next several hundred years, the evolution of people or changes in physical attributes will come more from science than from *chance* and *survival-of-the-fittest*. For example, good nutrition, vitamin pills, drugs and "new body parts" will make many changes in only a few years and the evolutionary changes will sort of be "lost in the shuffle." Of course, changes undoubtedly occur due to mixing of ethnicities and cultures, but these are short-term changes unlike usual longer-term evolutionary developments

Another way of saying this is: "***Thanks for your help, evolution. Now, we'll take over, thank you very much!***"

Evolution as the total explanation of human development

An "Experiment" relating to human evolutionary development

In order to more fully appreciate the reasoning behind the following discussion, it is suggested that those who wear glasses carefully clean them so that they are perfectly clear. If your vision isn't perfect, and you haven't had your eyes checked recently, you might get it checked and new glasses obtained.

CAN EVOLUTION TOTALLY EXPLAIN THE UNIVERSE MAKE UP?

Then, you should find a good seat, with a good view of many objects, preferably in the back of an auditorium filled with people or many objects in front of you. The ideal situation is in a place where there isn't a lot of action or distractions to take your attention away from what you are about to see. (Church, Synagogue or Mosque would be good.)

So, now you are quietly seated and looking at a number of objects in the near and far distances. Your eyes are wiped with a tissue and, if you wear glasses, they are adjusted and cleaned.

Now for the experiment:

Just gaze ahead and note how clearly you can perceive the differences in depth of the various objects. You can easily tell which objects are nearest to you and in the exact order as they progress in the distance, farther and farther away. Even things at quite a distance from you are clearly perceived relative to each other. *It's 3D without the movie glasses.* Not only that, but each object is in good proportion to the others - a short man and a tall woman in real life are also short and tall in comparison to each other, as are all objects in your entire field of view. Wow! You would pay plenty for this in a movie theatre and you get it every day, every month and every year, for *free*. What a deal!

OK, now cover only one eye and start over with your examination of this scene. Wait for just a minute until you become used to this new "mono-depth" view. Think about how it would be if you saw like this for all of your life. Think about how fortunate you are not to have lost the vision in one eye. You can't really tell much about the depth at all. You can guess about the relative position of each object and your mind can help out a little by knowing about the relative size of each object, etc., but you would be at almost a loss to distinguish the relative position of all of these objects.

THE UNIVERSE ISN'T JUST A BUNCH OF ROCKS

With one eye covered, you have lost most of your depth perception that you have always taken for granted all of your life (unless you have had vision problems, of course).

So, reflecting on this, you may start to appreciate the fact that you have two eyes and the accompanying brain setup to take advantage of them in order to give us this wonderful depth perception.

How did this happen that we got them?

Where or how did we get these two eyes? How come we don't have one or three or four? Funny, but all people normally have two eyes and almost all animals do too.

Is it an accident or evolution - chance and survival of the fittest?

The following is a simplified discussion of the science of evolution:

Is it an accident? An accident that all people and many different kinds of animals each have exactly, count 'em, two eyes? Let's see, if you assume that all people came from the same lineage and (in the *really* old days) the first animal accidently got two eyes, this would follow. But, wait a minute, we have two legs, two arms, opposing thumbs, eyelids that protect the eyeballs, etc. Lots of accidents, weren't there?

Or, is it evolution by survival of the fittest?

Scientists tell us that the first proto-eyes evolved among animals 500- to- 600- million years ago, about the time of the Cambrian explosion. Then, some say, evolution took care of the rest.

Animals probably evolved from marine protists, although no group of protists has been identified from an at-best sketchy fossil record for early animals. Multi-cellular animal fossils and burrows (presumably made by multi-cellular animals) first appear nearly

CAN EVOLUTION TOTALLY EXPLAIN THE UNIVERSE MAKE UP?

700 million years ago, during the late pre-Cambrian time. All known animal fossils during this period (known as the Vendian period) had soft body parts: no shells or hard parts, so they weren't able to be preserved as fossils.

The Cambrian explosion occurred, starting about 530-million years ago.

During the Cambrian period, animals evolved with external skeletons, hard body parts which allowed preservation as fossils. Due to this evidence of fossil records, it appeared that all the major body plans of animals arose in a short window of time.

Evolution theory points out the advantages of two eyes as well as two ears in the predatory survival "race." Two eyes are provided for depth and two ears for determining the direction of sound. At first thought, this sounds somewhat sensible in explaining the evolution of eyes and ears. But, wait! There is also the question of how the bodies of animals became bilaterally symmetrical. It appears that many animals, in our evolutionary line, have these types of bodies.

Now, the bilaterally symmetrical types of animals appear to have evolved before the sensory organs such as eyes and ears - or, perhaps at about the same time. Also, they evolved long before opposing thumbs and other things. It seems that the bilateral symmetry thing developed *before* the two eyes.

Bilateral *Symmetry is*, as you know, where both halves of the body are, roughly, mirror images of each other. Humans have bodies that are bilaterally symmetrical. This means that if you draw a line down the center of the human body, from the middle of the head to the feet, the same features appear on either side. We all have two eyes, one on either side of the body, and we also have two ears, two hands, two legs, and two feet.

THE UNIVERSE ISN'T JUST A BUNCH OF ROCKS

Then we have one of some of the *things in the middle*. We have one heart, etc. These middle parts aren't completely symmetrical, either.

Also, very importantly, the genes of bilaterally symmetrical animals control the development of "both halves" at about the same rate. Both eyes develop at about the same rate; fingers on each hand develop at the same rate. Even dimples, if the person is lucky enough to get one on each side, develop at about the same time.

At any rate, bilateral symmetry is the necessary ingredient providing the two eyes in animals. That is, with bilateral symmetry you *must* have two eyes (or an even number anyway).

Bilateral symmetry occurs in over 99% of animals.

The predatory advantage of this bilateral symmetry is not completely obvious, especially if you consider that this symmetry became quite prevalent very early in the development of organisms, including animals.

Of course, the advantage of being able to move rapidly has a clear advantage in survival of animals. Along with this mirror image development, bilaterally symmetrical animals usually had the sensory organs at one end. This obviously provided a predatory advantage to the animal.

CAN EVOLUTION TOTALLY EXPLAIN THE UNIVERSE MAKE UP?

An important thing to realize is that there is absolutely no advantage to left or right. There is, obviously, an advantage to having two of anything, but just two and not three or four or five? Why not two arms and four legs; or four of each? The Roman god Janus was always depicted with two faces, one facing forward and the other facing backwards. Now, here is the *mother of all predatory advantages*. Why didn't humans get this capability? We also didn't get keyboards on our stomachs, windshield wipers on our eyes, defrosters on our ears or, even internal genitals so we could all go around naked (and freeze to death). I might also suggest kidneys that "lasted 48 hours" and every girl with a switch that she could use to make her hair straight or curly. I could go on and on and I didn't even get to the *beer-can-pouch-built-in-for-males-only*.

In fact, the more that you consider it, there is a very big challenge to the idea that the total answer to human development is just chance followed by survival of the fittest (natural selection).

The challenge is: What happened to all of the things that we ***didn't*** get?

But, we did get depth perception. Isn't it lucky that we got the depth perception with bilateral symmetry? (It took the brain to help, of course, to actually get the depth perception.)

So, was the depth perception just a lucky outcome of bilateral symmetry? Would we have gotten two eyes and the depth perception without the bilateral symmetry?

(All of this keeps sounding like a *plan*.)

THE UNIVERSE ISN'T JUST A BUNCH OF ROCKS

The wide range of different levels of living things with their diverse groups of choices seems to indicate that evolution might be responsible for much of the variations. (For example, there are multiple eyes and other body parts in different organisms.) However, the whole evolutionary thing seems to be an extraordinarily simple answer to the question of design of our bodies.

It is also difficult to conceive as to how the evolution of human life was spread over the entire Earth after dry land appeared on the Earth. There weren't any pockets of developed animals on islands where different evolution occurred. The only answer to that must have been that all animals evolved and developed in one location and later spread around the globe. After all, there weren't many trains, subways, buses, airplanes and ships which were readily available to those lower animals; also, who would pilot these vehicles, furl the sails and steer? It might help a little to consider that much of this development occurred in the sea where there was better *transportation* available. However, all of the development didn't occur before animals left the sea for dry land. There is also the problem of "sea transport" to all points around the Earth.

There are a few other things that come into question when accepting the evolution theory as a *complete* explanation. Many of these arguments sound very much like the argument that infinity actually exists due to "something that we don't know."

It seems logical and reasonable that evolution has played a role in development of organisms including animals and, yes, to people too.

The question is whether it is the *total* answer to the development of everything up to, and including, humans.

CAN EVOLUTION TOTALLY EXPLAIN THE UNIVERSE MAKE UP?

The complexity of the human mind

One thing that needs to be considered when discussing human development through evolution, alone, is the complexity of the human mind. We don't understand at all the functioning of the brain and nervous system, but it is certainly an amazing and quite complex system.

The brain can be very abstract compared to the functioning of a computer but it uses functions that are really far advanced from any computer functions yet developed.

To start with, the brain doesn't use a binary system. It is theorized that it uses many levels of chemical states in each brain cell and element. Unlike computers, the human brain uses chemicals to transmit electrical signals. The brain also uses an assortment of vitamins and minerals to function.

Computers store data and the information doesn't go away unless the data is damaged or corrupted in some way. The human brain sometimes fails to store information, struggles to locate buried information, loses information and sometimes remembers things incorrectly.

The human brain adapts to new circumstances and learns new ideas. A computer can perform several tasks simultaneously. The brain manages many functions simultaneously, such as breathing, hearing, sight, etc. The computer can perform some calculations faster than the human brain, but the brain has the ability to come up with shortcuts and new ways of handling thought processes. The human brain can be imaginative and "think outside of the box" while a computer simply cannot!

One thing that should be remembered is that computers will continue to advance in speed and complexity in the future while the brain, in a reasonable time frame, won't see such advancement.

THE UNIVERSE ISN'T JUST A BUNCH OF ROCKS

However, there are simply a number of functions that the brain and the human nervous system performs that are quite different - and should be considered very advanced - over *any* computer.

The brain's remarkable functions in performing dreams

It is supposed that, in dreaming, the brain uses its "declarative memory" or its "episodic memory" system, which includes newly learned information.

The declarative memory stores information that you can "declare", such as 9 times 8, or the name of your dog. Sometimes, using episodic memory, you can even remember when or where you learned something - for example, when you first tasted pineapple upside-down cake.

Dreams also use "implicit memories." These are memories that are used in dreams even when individuals don't know that they have them. One type of implicit memory is found in memory where information is stored that you use without really being able to say how you know what you're doing. Examples are riding a bicycle for the first time in years, or typing on a keyboard without looking. Another type of memory used in dreams is one that involves general, abstract concepts.

As we all know, dreams may consist of illogically connected ideas or happenings in many cases.

One of the most interesting occurrences is the use of extensive "videos" in dreams. When a computer stores a video, it is stored exactly as recorded or edited and will play back exactly the same each time. Videos take a large amount of memory in computers.

In dreams, some very detailed long-running "videos" are played. These may sometimes have very detailed resolution with colors, smells, and even sounds as part of them. Some people dream of performing actions such as running, driving or flying.

CAN EVOLUTION TOTALLY EXPLAIN THE UNIVERSE MAKE UP?

Included in these dreams are all kinds of details and plots. However, these videos are sometimes under the control of the individual doing the dreaming.

A personal dream experience

As a personal experience, when I was in my late teens, I decided that it would be interesting and fun if I could "know" when I was dreaming. The idea was that I could then do all kinds of fun things like jumping off of buildings to confound my friends. I could also do things which were wrong in real life and that I would never do, but I could do while dreaming - such things as rob a bank, beat up someone, and perform female conquests at will - without ever actually doing anything wrong. I would not have to answer to the authorities or have a guilty conscience. And, I would not have to worry about God and any consequences. Wow, what a deal! The problem was, "How would I be able to be sure that I was dreaming?" I had to be absolutely sure or there would be Hell to pay.

I did manage to do it. I determined that the only way to be absolutely sure that I was dreaming was this: If I wasn't sure, then I was dreaming. It was foolproof. If I was awake, I was sure that I was awake. If I was dreaming, I wasn't absolutely sure, so I WAS DREAMING. I could take it to the bank and I did.

Now, unfortunately, things didn't work out much of the time. When I jumped off of a building, my friends weren't awestruck. When I did some flying around, they hardly noticed. When I robbed a store, I had some money, but so what? I couldn't spend it in any of the lavish ways that I had visualized. Finally my conquests just didn't work out at all at the "conquest-ing" point.

Dreams are a really powerful capability

The point of telling this story of my personal dream experience is to emphasize the extraordinary functions of the brain in "live video playing."

It is almost unthinkable that a computer could even approach these functions. The brain is actually creating an interactive video in real time (approximately) with all of the surrounding details, colors, sounds, and other effects. Finally, upon awakening, the "video" can be, at least partially, replayed over and over again.

This whole dream capability seems to extend the capability of the brain. The evolution, solely through chance and natural selection, seems less likely.

The final knock-out-blow against "chance-only" human development

As we say: "Now for the *Knock-Out-Blow* against the theory that evolution is *totally* responsible for human development."

Here it is:

Remember, as previously stated, evolution of the *Universe,* including space, matter, time, **and** the rules, does not follow the ideas of "chance" and "survival of the fittest."

If there was a planner involved in creating the Universe, there is no question that the same planner and creator also worked in the development of humans.

I think that logic points to human evolution being both guided **and** using natural selection.

10 What about the Universe Equations?

I have discussed the makeup of the Universe, including matter, waves, rules/laws and all of the other elements.

One interesting fact is that many equations and fixed numbers are included in the rules.

Why do we have the equations in our Universe that we do?

Why do time and space relate as they do?

It is easy to see that some equations fit our reasoning, like the area of a rectangle being one side times the other side. The area of a circle being pi times the radius, squared (ratio of a circle's circumference to its diameter). This makes some sense, since it must be less than 4 but I have always wondered why the Universe rules settled on 3.14159265.... etc. as the factor when, although it is certainly proven to be correct, there is no way to understand this factor based on logic.

It is also understandable that gravity or magnetic attraction should decrease with the square of the distance apart. This makes some sense since there is an area (square) rather than just dividing.

Many of the equations are not quite so easy to understand, like Ohm's law:

$V = I \times R$ or Voltage (V) = Current (I) multiplied by Resistance (R)

Or, Newton's three laws of motion:

1. An object that is at rest will stay at rest unless an unbalanced force acts upon it. Also, an object that is in motion will not change its velocity unless an unbalanced force acts upon it.

THE UNIVERSE ISN'T JUST A BUNCH OF ROCKS

2. The acceleration **a** of a body is parallel and directly proportional to the force **F** and inversely proportional to the mass **m**, i.e., **F = ma**.
3. To every action there is always an equal and opposite reaction: or the forces of two bodies on each other are always equal and are directed in opposite directions.

Or, the Lorentz transformation time dilation factor:

$$t = t'/\sqrt{(1-v^2/c^2)} \quad \text{or} \quad t = \frac{t'}{\sqrt{1-\frac{v^2}{c^2}}}$$

Where: t' is the passage of time measured by a stationary observer.
t is the passage of time measured by an observer traveling at velocity with respect to the clock.
v is the velocity.
c is the speed of light.

The big question is this:

How did the equations come into being?

Was there a master Planner/Creator involved?

I can almost see a mathematician at work - a really **good** one, to say the least.

11 Pause for a Moment, Just for Emily

A while ago at a family gathering, we were discussing my previous book, *If I Had Been Born a Muslim*. This book was about God, the Creator, Religions and such. I noted that I was working on another book at which point, my teenage granddaughter, Emily, said "Grandpa, when you write another book, you should write one that people would like to read." Then, she hastily added, "Like a romance novel."

Well, Emily, I am including a couple of stories, as you requested.

Sarah and the Lost Scarf

One day, Sarah and her two good friends, Mary and Jenni, all in their early twenties, were walking in the woods. It was autumn and there was a slight chill so Sarah had wrapped her favorite, brightly colored scarf around her shoulders. As they were walking, they noticed a nice looking fellow a short distance away. Sarah couldn't help noticing that, as he walked, the scattering bits of sunlight shimmered on his rippling muscles. She turned quickly away so that he didn't notice her staring. As Sarah stooped to pick some flowers, the end of her scarf fell across a branch. Suddenly, a squirrel darted across her path and snatched the end of the scarf and sped away into the woods. The man, noticing what had happened, dashed after the squirrel through the woods at top speed.

As Sarah saw it; *He climbed onto his trusty steed and, totally casting aside any thoughts of his own safety, sped after the dragon. Evidently, a sorcerer threw up a giant wall of flame and the knight was totally engulfed. But, somehow, he escaped and returned. After dismounting from his trusty steed, and removing his armor,* he came up to Sarah and, out of breath, said "I'm afraid that he got away."

Sarah sighed, "Oh, thank you so much for trying. I will be forever grateful for that."

THE UNIVERSE ISN'T JUST A BUNCH OF ROCKS

Lance lightly touched her shoulder with a finger and said, "You must be cold without your scarf."

Sarah, gazing into his eyes, shuddered with his touch but composed herself, quickly and said, "It's good to meet you, and I'm Sarah".

He responded, "I'm Lance. Maybe, just in case that I see the scarf, you should give me your cell number?"

Sarah quickly gave him her number and said, "I hope that you will use it even if you don't find the scarf."

He replied, "I certainly will, and soon."

As he walked off, Mary and Jenni came up and Mary said, "What a good looking guy. You seemed to think so too."

Jenni asked, "What is his name? Did you like him?"

Sarah sighed to the depths of her being, smiled and said, "His name is Lance and yes, I really liked **Lance a lot.**

PAUSE FOR A MOMENT, JUST FOR EMILY

Julie and Kristen have a narrow escape

Friends, Julie, and Kristen were discussing the start of school at their middle schools as they walked their dogs, Holly and Jazzy. They were so engrossed in discussions that they almost failed to notice the municipal bus that was approaching from behind. (As a note, I should tell you that riding on that bus was the famous, nationally-known sportscaster, Ally Wilhelms, who was on her way to handle a San Francisco Giants baseball game.)

As the two girls walked and talked (mostly talking) the bus driver had his window down and, suddenly, without warning, a bee came through the window and landed on the end of his nose. He swatted desperately at the bee with one hand while trying to control the bus with the other. The bee took that as an affront and responded by giving everything that he had into one giant stab with his stinger right into the end of the bus driver's proboscis. With the driver distracted, the bus swerved and ran right up on the sidewalk and, *get this*, directly at the two girls who were still walking their dogs, obliviously.

Well, before you get too worried, let me tell you that this story has a happy ending!

So, what happened is this: One of the two dogs reacted. (I won't say which one because I don't want to show favorites.) This hero dog turned and ran directly at the two girls and threw herself directly at both of them at once, knocking them out of the way of the bus - just in the very last nick of time. Ally recorded the hero dog's actions on her video camera to show on TV and the dog was nationally recognized. A group representing the Giants' Public Relations, namely Peter, Dan, Tom, Joe, Jackson and Ryan, awarded the dog a trophy in an on-field ceremony.

One more thing: The other dog was very narrowly missed by the bus but did have a slight mishap. As the bus roared by, it ran ignominiously, over her tail. I'll let the dog finish my story for me:

THE UNIVERSE ISN'T JUST A BUNCH OF ROCKS

The other dog said: "Well, that's the end of the *tale*".

12 Are Time Machines Possible in The Universe?

The answer is unequivocally, "YES," and I can prove it!

But, NOT the science-fiction type of which you are thinking.

Instead, a REAL one! One that really works!

Just wait for the exciting ending to this story:

A look at the problem

Many interesting science-fiction stories have been written about time machines. In these stories, a person goes from one time period to another at the touch of a button and then returns to his original time period, again using his time machine.

A realistic technical look at possibilities for this type of time travel is, of course, discouraging.

The question is this: Can any possible technology be evolved that will provide us with the capability to alter time so that we can appear in another time period?

Then, more importantly, can we return to our time or to another time period.

If I examine the latest technology, according to the rules of Special Relativity and General Relativity, I am presented with faint possibilities. For example, I could alter a clock speed/rate, in comparison to another clock, through very high velocities. Also, the time *rate* can be adjusted through exposure to extreme gravities.

These adjustments to our clocks relative to another clock would only make the different clocks run, one slower or faster, *relative* to the other.

Such unknowns as "black holes" present interesting possibilities for adjusting our "clock rates" when compared to other clocks.

These are just theoretical possibilities since the extreme accelerations necessary to get to very high velocities (approaching the speed of light) would wreak havoc with the human body - to say the least. They also require huge amounts of propulsion to obtain these velocities. The exposure of clock *slowdowns* due to a close proximity to a "black hole" would be extremely destructive to any matter and especially to the human body.

There are strictly hypothetical "bridges" between regions of space called "wormholes." These would (theoretically) allow "jumping" from one region of space to another and might allow for "jumping" to another time period. Scientists have theorized that negative energy that they call "exotic matter" would be required to maintain these wormholes. Of course, no "exotic matter" has ever been created.

These theories are just mathematical formulations with only very faint hopes of any success. However, this kind of future technological advance cannot be discarded - with a 100% certainty.

While time is certainly a part of the Universe equations, all of the scientific knowledge - at this point - indicates that time travel of this type is impossible and will remain so in the future.

While it appears that time travel involving reversing of time is absolutely impossible, as eternal optimists we should keep an open mind but, at the same time, basing our actions and our lives on a conviction that such reversing won't ever happen.

ARE TIME MACHINES POSSIBLE IN THE UNIVERSE?

The *answer* about time traveling

Just using time dilation to adjust clocks for two observers won't really allow time travel as we desire. That is, traveling backwards in time.

Future technological advances offer only faint hope. Possibilities that physics might allow time travel are only faint, one in a billion *or more* chances.

However, there is another problem that may absolutely mean that the Universe rules will disallow backward travel in time. This is the problem of altering the future when going back in time. This has had much discussion through the years with things like the "grandfather paradox." In this supposed paradox, you build a time machine. Then, you travel back in time, meet your grandfather before he produces any children (i.e. your father/mother) and kill him. Thus, you would not have been born and the time machine would not have been built, a paradox. Of course, there is no such thing as a real paradox. This statement means that it couldn't have happened.

The previous discussion concerned a *possible* paradox. However, if there is Free Will, or any non-deterministic actions in the Universe, any appearance in a previous time would cause an absolutely *certain* paradox. The future would be different since it would be changed just by the *appearance* of someone in this previous time period. Then, the time traveler could never go back to his time from which he came since it wouldn't exist anymore in its original form.

Therefore, if there is Free Will (or any non-deterministic action) there can't be time travel in the Universe - it can't be allowed. Any backward time travel would cause a paradoxical situation and real paradoxes aren't allowed in the Universe. I must assume that non-deterministic actions do occur. (See later discussions.)

A possible exception to this rule:

Time travel to the past would (theoretically) be allowed if the time traveler were not allowed to be seen or to interact in any way with the past existence. This means that the time traveler could exist without form in the past. He could see and observe things; but could not in any way cause change in the past existence.

This, of course, is very theoretical but does offer the only possibility to remove the impossible paradox.

Let's examine all of these things in the following.

Clock-rate adjustment to adjust time?

Special relativity

Special relativity is the physical theory of measurement in inertial frames of reference proposed in 1905 by Albert Einstein (following the independent contributions of Hendrik Lorentz, Henri Poincaré and others) in the paper "On the Electrodynamics of Moving Bodies." It generalizes Galileo's principle of relativity - that all uniform motion is relative, and that there is no absolute and well-defined state of rest for either of the bodies in motion.

Special relativity also incorporates the principle that the speed of light is the same for all inertial observers regardless of the state of motion of the source.

This theory has many consequences which have been experimentally verified, including ones such as length contraction and time dilation. The effect has been verified experimentally using measurements of precise clocks flown in airplanes and satellites. This contradicts the classical notion that the duration of the time interval between two events is equal for all observers.

ARE TIME MACHINES POSSIBLE IN THE UNIVERSE?

Einstein's twin paradox

The "twin paradox" is a thought experiment in Special Relativity by Einstein, first done about 100 years ago.

In this thought experiment, a twin makes a journey into space in a high-speed rocket and returns home to find he has aged less than his identical twin who stayed on Earth. According to the special theory of relativity, time on the travelling space ship, which is travelling at speeds approaching the speed of light, runs slower when observed by a stationary reference source, such as one on Earth.

This result appears puzzling because each twin sees the other twin as traveling, and so, according to the theory of time dilation by Special Relativity, each should paradoxically find the other to have aged more slowly.

This is what is referred to as the twin paradox.

There have been numerous explanations of this paradox, many based upon there being no contradiction because there is no symmetry --- only one twin has undergone acceleration and deceleration, thus differentiating the two cases. One version of the asymmetry argument is that the traveling twin uses two inertial frames: one on the way out and the other on the way back. So switching frames is the cause of the difference, not the acceleration.

As an example, consider a space ship traveling from Earth to the nearest star system outside of our solar system: a distance of 4.45 light years away, at a speed of 0.866c, that is 86.6 percent of the speed of light. The Earth-based observer reasons about the journey this way: the round trip will take time = 2 times the distance / velocity = 10.28 years in Earth time (everybody on Earth will be 10.28 years older when the ship returns). The amount of time as measured on the ship's clocks and the aging of the travelers during their trip will be reduced by the factor ش = $\sqrt{(1-v^2/c^2)}$, which is the reciprocal of the Lorentz factor. In this case ش = 0.500 and the travelers will have aged only 0.500×10.28 = 5.14 years when they return.

The ship's crew members also calculate the particulars of their trip from their perspective. The travelers' result is in complete agreement with the calculations of those on Earth, though they experience the trip quite differently.

If a pair of twins is born on the day the ship leaves, and one goes on the journey while the other stays on Earth, they will meet again when the traveler is 5.14 years old and the stay-at-home twin is 10.28 years old.

A paradox in logical and scientific usage refers to results which are inherently contradictory, that is, logically impossible.

Actually, the twin paradox experiments are carried out every day in the GPS satellites. The experimental results show that we see "old" and "young" atomic clock "twins." The results predicted by general and special relativity theory are correct, however, the interpretation of relativity gives rise to the paradox. We can claim that the clock in the GPS satellite underwent motion relative to the clock on Earth, as well as the other way around.

The Global Positioning System (GPS)

The current GPS configuration consists of a network of 24 satellites in high orbits around the Earth. Each satellite in the GPS constellation orbits at an altitude of about 20,000 km from the ground, and has an orbital speed of about 14,000 km/hour. The orbital period is roughly 12 hours. The satellite orbits are distributed so that at least four satellites are always visible from any point on the Earth at any given instant (with up to 12 visible at one time). Each satellite carries with it an atomic clock that "ticks" with an accuracy of one nanosecond (one billionth of a second). A

ARE TIME MACHINES POSSIBLE IN THE UNIVERSE?

GPS receiver determines its current position and heading by comparing the time signals it receives from a number of the GPS satellites (usually 6 to 12) and triangulating on the known positions of each satellite. Even a simple handheld GPS receiver can determine your *absolute* position on the surface of the Earth to within 5-10 meters in only a few seconds. A GPS receiver in a car can give accurate readings of position, speed, and heading in real time.

To achieve this level of precision, the clock ticks from the GPS satellites must be known to an accuracy of 20-30 nanoseconds. However, because the satellites are constantly moving relative to observers on the Earth, effects predicted by the Special and General theories of Relativity must be taken into account to achieve the desired 20-30 nanosecond accuracy.

Because an observer on the ground sees the satellites in motion relative to them, Special Relativity predicts that we should see their clocks ticking more slowly. Special Relativity predicts that the onboard atomic clocks on the satellites run slower than clocks on the ground by about 7.4 microseconds per day because of the slower ticking rate due to the time dilation effect of their relative motion. (Of course, the basic ideas of Special Relativity will indicate that the **opposite** should occur, since either the Earth or the satellite can be considered "at rest" in inertial frames.)

A prediction of General Relativity is that clocks closer to a massive object will seem to tick more slowly than those located farther away because of gravity. A calculation using General Relativity predicts that the clocks in each GPS satellite should be running faster than the ground-based clocks by about 45.7 microseconds per day.

The combination of these two relativistic effects means that the clocks onboard each satellite should tick faster than identical clocks on the ground by about 38 microseconds per day (45-7=38).

THE UNIVERSE ISN'T JUST A BUNCH OF ROCKS

To make up for this difference, the clocks and the computers in the satellites are adjusted by the 38-microsecond factor.

What do you know! It works! If it didn't, the GPS system wouldn't work, of course.

Now, for the big question:

The Earth is considered as *the reference frame "at rest."* There is no reciprocal arrangement of reference frames "at rest" which Special Relativity predicts for inertial frames. If there was, the clocks in the satellites would run **slower** than the clocks on the Earth.

Why?

There are three **differences** from the GPS satellites and the Special Relativity reciprocal time dilation theory due to velocity of two inertial frames.

1. The acceleration used to get the satellite into orbit.
2. The gravitational force on the satellite.
3. The centrifugal force (due to its velocity) counteracting the force of gravity that keeps the satellite in orbit.

Comparing the twin paradox and the GPS system

It is proven by the GPS-system operation that the time dilation due to the velocity (Special Relativity) is correct and that the reference **frame at rest** is set as the **Earth**.

It is obvious that one (or more) of the factors listed above was the reason.

Perhaps, it can't be proven conclusively exactly which factor is the explanation, but the results confirm that the Earth frame of

ARE TIME MACHINES POSSIBLE IN THE UNIVERSE?

reference is "at rest."

The factors in the Twin experiment include acceleration and deceleration and non-inertial frames of reference.

Is the reference frame "at rest" set as the one on the Earth (the one that the twin left behind)?

It seems to be clear that it **is** since the same acceleration away from the Earth (plus the other forces) is present as in the GPS situation.

It isn't possible to have both the "travelling twin" and the "stay at home twin" to be the younger.

The conclusion that has to be reached is that the Earth frame of reference will be the one which is "at rest" in both situations.

What is a paradox?

To again emphasize a point: As there is never an actual infinity, there is never an actual paradox, since this is a violation of the definition in either case.

The word, paradox, is used to describe contradictory conclusions. A contradiction of logic isn't allowed in reasoned conclusions.

At present, the twin experiment isn't achievable.

There are many problems with actually performing the "Twin Experiment" in reality. These center on the huge amounts of propulsive energy necessary and the potential damage to a human due to the tremendous acceleration and deceleration forces involved in reaching velocities approaching the speed of light.

So, while this kind of "clock-rate adjustment" isn't presently achievable, the future possibility shouldn't be completely discarded.

In addition, the advantages that are provided by performing this clock-rate adjustment thing aren't really obvious. Certainly nothing like the desired "Time Machine" goals is provided.

So, let's discard this approach and look for another.

First, let's look at some questions which have an important bearing on our quest.

Is there free will in the Universe?

The question of free will or non-deterministic actions in the Universe is an important one. It is also a key to examining any possibility of "Time Travel" in the future.

To begin with, I must accept that it is very difficult to *prove* whether or not there is any free will or any non-deterministic actions. All I can do is to lay the cards on the table and try to provide details of what is involved in the options and what difference these things make.

I will try to use plain and fairly simple language in these discussions.

Free will and deterministic actions

Free will refers to the freedom of humans to make choices that are not determined by prior causes (determinism) or by divine intervention.

Determinism is a theory that acts of the will, occurrences in nature, or social or psychological phenomena are totally determined by preceding events or natural laws.

ARE TIME MACHINES POSSIBLE IN THE UNIVERSE?

Is there randomness in the Universe?

There have been many arguments throughout history concerning this subject, using complicated language. Let's try to make it simple.

Most activities which are described as random acts are, in fact, deterministic. Another way to say this is to say that randomness is really uncertainty, but the outcome is a certainty given that all of the contributing factors stay the same.

As an example, consider the roulette wheel in the casino game of roulette. A ball is spun around the wheel while the wheel is rotating in the opposite direction. After usually 6 or 7 revolutions, the ball falls onto the wheel and bounces around before finally coming to rest in one of the 38 slots on the wheel. The ball's travel in reaching its final destination seems to be random, but it's really not. There are just so many factors which go into the determination of which slot the ball finally falls into that it seems almost random. Factors such as wheel speed, the speed of the ball being thrown against it, the topology of the surfaces, the air density, any wind or vibrations - all account for the determination of the result. When the ball hits against the wheel surface, it bounces around, again looking as though it's random but it is all deterministic and if you could calculate all of the variations of conditions, you could, in fact, calculate into which slot the ball would fall.

Most games involving seeming randomness involve the same situation; such complicated conditions that it is really, really difficult to determine the outcome. It's all the same with rolling dice in a craps game, flipping a coin to see if it lands heads or tails, picking a slip of paper out of a bag to see what number is drawn, etc. It is all deterministic but it is extremely difficult to predict the result in advance.

THE UNIVERSE ISN'T JUST A BUNCH OF ROCKS

Even computer-generated *random* numbers aren't really random at all. Often, for example, a mathematical equation is used in the computer program, with a result which seems to have successive integers which are so mixed and unpredictable that it seems (and may, actually be) *random*. An interesting thing with computer-generated random numbers, though, is that the computer will come up with the same series of numbers if the program is run again.

The usual way to fix this problem is to start with a seed number fed into the equation, so that the series will be different with each different seed. A common method of determining a seed is to use the number of fractions of a second in the current clock time as a starting point.

Now, even these aren't random but they are so varied and unpredictable, and use the same predicted probabilities that the results approach those of random numbers.

So, there aren't really random numbers that are usually used in all cases where randomness is desired - just numbers which are varied, unpredictable and in all aspects, with the same *probabilities*.

There are people who argue that there is really randomness in the Universe. They argue that quantum physics shows that the behavior of extremely small particles use probabilities rather than deterministic certainty. This is partly based on the difficulty of observation since an electron or a photon used to try to measure action, causes reaction in and of itself. But, the scientists argue that mathematics prove that possibility exists.

ARE TIME MACHINES POSSIBLE IN THE UNIVERSE?

What is the conclusion about randomness in the Universe?

It seems that there are just two possibilities for random or non-deterministic outcomes in the Universe:

1. There may possibly be random things going on in the physical world, but it appears difficult to prove this right now. It is fairly safe to guess that things are *usually* deterministic. This is important in the discussions of possible future Time Travel using advanced technology. It may be that a portion of the Universe rules does provide random actions of particles at the quantum level.

2. There is the possibility that people have free will in their decision-making abilities. This would mean that they are capable of non-deterministic decisions and actions. (It must be stated that it is even possible that lower-level animals also possess this capability.)

Use future technological discoveries to provide time travel.

We have to be always on the lookout for a new discovery to let us have time travel - but we shouldn't hold our breaths! Many theoretical possibilities are suggested (some with faint possibility of success), such as:

> Transversable Wormholes
> An Alcubierre Drive
> A Cauchy horizon
> Traversable Wormholes
> A Tipler cylinder

As previously stated, these possibilities shouldn't be completely discarded since such technological methods can't actually be *proven* to be *impossible*.

THE UNIVERSE ISN'T JUST A BUNCH OF ROCKS

Proof (99.99999999% - or more) that real, physical time-travel is impossible.

As previously discussed, if the concept of free will is allowed or, if any non-deterministic (random) actions are allowed, there is no predestination. In other words, if the same historical scene were to be played multiple times, there would be multiple outcomes. What this says - it guarantees it - is that it would not be possible to come from the future and then return to the same future. This is because the future would no longer exist as it was before. This isn't the same as talking about altering the future through some actions of the time traveler, while in the past. It is saying that there would *always* be some altering of the path if progression to the future is repeated.

If you don't accept free will, then everything is preplanned and our lives are basically worthless here on Earth because we don't really make decisions, we just play out the predestined course. Note: Our actions aren't good or bad, they are just the "next part of the movie." This is rarely accepted by anyone since it takes away much of the importance and meaning of life.

There is another word which comes out often, when someone is discussing free will. This is the supposed *omniscience* of God. It is sometimes stated that God, being omniscient, knows everything which will happen. Then, it may be stated that, since God *knows* what will happen, there is no real free will. (Again, if God *knows* what will happen, the outcome of the "story of the future" does not allow free will.)

Now, I'm going to try to get into some really hot water:

The supposed omniscience of God is an absolutely ridiculous concept!

Before you start to stone me for being heretical, let me add that what I am stating is simply a logical statement.

ARE TIME MACHINES POSSIBLE IN THE UNIVERSE?

The definition of omniscience is: Knowing everything - having an infinite knowledge or understanding.

Reaching infinity is, of course, against the definition. So, the fact that God is not omniscient, is not because he isn't powerful enough, it is just because that is against the definition of omniscience.

The only way that omniscience would be possible, according to logic, is that the "everything" in the definition is *bounded*. That is, that there are a finite number of things about which to have knowledge. And, that seems to be impossible.

So, to sum up:

God may know an awful lot. He may know millions, billions and lotsmorillions of things. He could even know much of what will happen in the future - in theory, but **not**, in all probability - absolutely everything.

My own supposition is that everything about God's plan and recorded statements by prophets galore from the past, says that he set things up with **free will** as part of his plan.

At least, that's what I plan to go on in the future, since to discard free will makes my part fairly insignificant (even more so in the future than now).

If we have no free will, then time machines don't seem very interesting - nor does much of anything else, either. I hate to say it but that cute blonde sitting at the next table wouldn't fascinate me much either - well, maybe just a little - with no free will. (However, not nearly as much as if I thought that she was **free** to do as she **willed**.)

THE UNIVERSE ISN'T JUST A BUNCH OF ROCKS

Nobody from the future has contacted us.

There is another very strong proof, or at least a very strong indication that time travel is not technically possible.

This is simply that nobody from the future has yet contacted us.

If the rules/technology of the Universe would allow time travel, I could feel sure that someone from somewhere in the future would have discovered it and would have visited our time. Now, it is *possible* that they would have kept it a secret from us and not allowed the news to appear on the front page of *The New York Times*. Time travelers could have used a mandate that they must not alter their past. This would make sense, but it doesn't seem reasonable that the Universe rules would *allow* time travel but not allow for or compensate for the problem of "Past Altering."

Let me say that again, in order to make it clearer:

The Universe rules have been set up in a fantastically well-coordinated, integrated, everything-fits-and-works-together-well way. If time travel were physically allowed by the Universe rules, it seems incongruous that the Universe rules wouldn't have been set up to take care of the problem of going back into time and altering things which would then royally foul up the future in the process. The Universe rules are so well set up in other cases that this problem wouldn't have been left to humans setting stiff rules against *past altering*.

All in all, I feel that, if time travel were technically possible, someone from the future would have contacted us and their presence would have been made known - at least to someone.

So, although changing the relative clock rates of different observers has been proven, my vote for actual time travel is: Physically moving backward through time isn't possible.

ARE TIME MACHINES POSSIBLE IN THE UNIVERSE?

Note: Many people, including technically competent scientists have tried to come up with theories which would explain away the previously described compelling arguments against time travel. These have required very unusual ideas such as parallel universes and other fairly outlandish concepts. None of these theories seem to be based on real science but on fairly unscientific, not logically cohesive, conjecture.

Use "Virtual Reality."

How do you attack problems like time travel, which apparently have no solutions?

A nationally published columnist (Marilyn vos Savant) recently stated in response to questions from a reader, that many, many things were impossible. The columnist made this statement because she is a technical scientific mathematical (and very competent, I will add) type of expert.

From a philosophical viewpoint (as a philosopher), it may be stated that many, many things are possible. Certainly, it is valuable to start with the philosophical assumption that "anything is possible."

When something appears impossible, a good approach is to back off, regroup and attack the problem from another angle.

This approach seemed to work many times, especially in my work on our computerized Hospital Information System, which we developed in the 1970s and 1980s. When something appears impossible, go at it from another angle. People will accuse you of cheating. They will say, "That's not fair." Often though, you can realize your goals.

It is important to be able to realize the huge possibilities of future development in computers and in computer application software in order to be able to see the possibilities in the future.

THE UNIVERSE ISN'T JUST A BUNCH OF ROCKS

A later section in this book covers some of the exciting forecasts of computer and advanced development of all kinds.

The approach of achieving time travel through the use of virtual reality is to take advantage of two things:

1. **The future advances of computer and related technologies.**

2. **The virtual world of virtual reality - not a science-fiction story but an achievable reality.**

So, **virtual time travel to the past** will take advantage of the knowledge of history PLUS the interpolation and extrapolation of information by advanced computer applications. Libraries of books and written materials placed online will be scoured by the computer applications. Complete stories together with videos will be made available to the users through the "Virtual World."

The real possible accomplishments in this area are realizable when you get some glimpses into the possibilities of advanced development utilizing computers. Just think of what has already been achieved in Internet browsing. You can type a question and feed it to one of the search engines. The results seem amazing when the question is either answered directly or you are immediately pointed to answer sources. Well, this is only the barest of beginnings. You ain't seen nothin' yet! Capabilities are doubling each year or two. Thousands and thousands of hours are constantly spent working on search engines. It doesn't take much imagination to visualize what will undoubtedly take place with future advances.

Virtual time travel to the future will take advantage of history but will use the computer prediction applications to make educated and highly accurate predictions of future "histories."

ARE TIME MACHINES POSSIBLE IN THE UNIVERSE?

It should be noted that predictions of future development, which specifically are not too accurate in details, are often surprisingly accurate - *in a general sense.*

The future applications may take longer to develop than previous history compilations. However, I am talking about *maybe fifty or a hundred* years, not *five hundred.*

See descriptions of future technological advances and of virtual reality in later sections of this book.

THE UNIVERSE ISN'T JUST A BUNCH OF ROCKS

13 The Wildly Fantastic Future Because of Computers and Advanced Technology

A general thought

To try to get things into perspective about the future, let's look briefly at the past.

In many ways, technology has increased at an accelerated rate in the past.

So look at the progress (simplified) in periods (I'll use BCE and CE rather than BC and AD):

4000 BCE: We got wheels.

1000 BCE to 0 CE: The world was thought to be flat. No engines, or electricity. There were metal weapons, etc. We got catapults and finally, horseshoes.

1 CE to 1000 CE: World still flat, no engines or electricity. We still have metal weapons. Better ships with sails, now.

1001 CE to 1492 CE: The "flat" world is "getting round," no engines or electricity. We've got gunpowder but only simple guns. Very much the same ships.

1493 CE to 1800 CE: The world is really round now and finally there are crude steam engines, no electricity. Guns are a tiny bit better.

1801 CE to 1870 CE: Found out how far it is around the world. Good steam engines and steam ships, no electricity. We have pretty good guns now. Horses are ridden a lot.

1871 CE to 1900 CE: We've got the first light bulbs. The Industrial Revolution is here. Telegraphs and trains abound.

THE UNIVERSE ISN'T JUST A BUNCH OF ROCKS

1900 CE to 1945 CE: First cars, airplanes, telephones abound. We've got lots of guns and bombs. We made atomic bombs at the end of this period (maybe we're not too proud of that).

1945 CE to 1999 CE: Computers get started. Internet takes off toward the end of the period. Rockets and satellites soar. We even go to the moon. Rudiments of Artificial Intelligence begin in earnest. Little and larger wars use "fancy" aircraft and weapons.

2000 CE to Now: Lots of electronics, Internet is used by everybody, cell phones, smart phones, high-def TV. Cars get pretty reliable and use increased technology including computers.

These simplified statements of technological advancement are just shown to demonstrate that technology has been increasing at an exponential rate. This means that it is increasing faster, every year, every month, and every day.

There is every reason to assume that this accelerated advance will continue in the future.

The main message, here, about the future is, simply **WOW!**

History of computer-power evolution

Almost 40 years ago, Gordon Moore of Intel observed that the transistor density doubled every two years or so. This became known as "Moore's Law." Actually, there are far more general observations and their corollaries which are possible and lead to profound conclusions. These conclusions are that computer power, communications speeds and capabilities and other related functions **have, and always will, increase at remarkable rates**.

THE WILDLY FANTASTIC FUTURE BECAUSE OF COMPUTERS AND ADVANCED TECNOLOGY

Let's start at more than 50 years ago with an examination of computer development in the late 1950s. At that time only computers which I'll call *main frame computers* were available. In a few years, about 1960, *mini computers* were available. Although *personal computers* were technically available starting in about 1969, they really didn't start to gain momentum until about 1973 with Apple and IBM computers. Let's call another class of computers, *server computers*, which really started to gain momentum after Sun Microsystems was formed in 1982. At this time, mini computers sort of merged and became *server computers*.

The numbers following present the **rates** at which computer capability **doubled**. **B**asically all of the associated technologies should be included in these doubling rates. When I say computer capability, it can be stated in many terms. Factors include computer or CPU speed, main or RAM storage, auxiliary or disk storage speed and size and removable disk or media size and speed. Communications speed is an important factor. Included are even printing quality or density, display quality or density, sound input and output, picture and video resolution, software sophistication, etc.

Now some things doubled in fits and starts, but over a period of years, the average rates stayed very consistent. For example, communications speed started with 150 bits per second or bps. Next 150 bps, 300bps, 600 bps, 1200 bps, 2400, bps, 4800 bps, and on and on to 56,000 bps, then to higher speed "DSL" and now, "the sky's the limit."

THE UNIVERSE ISN'T JUST A BUNCH OF ROCKS

Display screens have progressed fantastically as everyone knows. At first, characters on computer screens were displayed using a 5 by 7 dot matrix. Then characters were displayed in increasing resolution. Then, multiple fonts were displayed, followed by graphics including pictures, then video. Televisions went from grainy 512 lines of dots displayed in black and white on cathode ray tubes to color (in about 1960) to high definition displays including Blue Ray DVDs. Next will be holographic display and then .....

Now, for computers, the numbers following are presented not as technical numbers, exactly correct to several decimal points. They are presented as philosophical numbers which are basically correct, based upon years of observation and which lead to my conclusions about the past and the future.

Main frame computers were first introduced in about 1955, but, let's start in 1958 when computers started doubling in capability at a rate of about every three years. Perhaps in the late 1960s this increased to about every 2½ years, then, this rate stayed there until the 1970s where it dropped to maybe 2¼ years.

Personal computer capabilities started doubling about every 2½ years. About 1975 or so, this dropped to about every 2 years, or 24 months. In the 1980s it dropped to 22 months, then 20 months, then 18 months and finally about every 15 months. At the peak of the doubling rate, for a few years it approached **every 13 months**. This appears to have slowed at present to about every 15 to 18 months.

Server computers have doubled at somewhere between the main frame rate and the personal computer rate.

What the above means is that all of these computer capabilities have consistently doubled at a rate well under every two years and probably at an average rate of about every 1½ years.

124

THE WILDLY FANTASTIC FUTURE BECAUSE OF COMPUTERS AND ADVANCED TECNOLOGY

As computer capabilities increased, there have been steps in computer applications which have "eaten" much of the advances. First, characters were formed using dots on the computer screen. Then, graphics required a lot more computer capacity. After graphics, sound and pictures required more capacity. Then video, including television "ate up" more. Then the Internet and high-definition television gobbled and gobbled.

Another thing has also happened. I call it "Object Development." This refers to a computer development which somewhat parallels the Universe plans. In both of these, objects perform functions more or less independently and can be called on or utilized by anybody. We are all familiar with the fantastic advance of search engines and "apps" (applications) which can be used on the Internet, or downloaded to personal devices. The secret of rapid development in "apps" is important. Many thousands of people can get involved in developing, causing even faster advances.

Future of computer-power evolution

I have stated that computer capabilities would increase indefinitely since computers aren't **real** but are only a simulation. The statement that they aren't real but are only a simulation isn't completely accurate. For example, there are communications speeds as well as display density, printer density, voice processing, etc. However, the basic truth is correct. Computers can continue to advance and advance, and it is reasonable to assume that this will go on and on ad infinitum.

Note: The ad infinitum was only an expression since, as you know, if you have a brain in your head, infinitum isn't possible just like infinity isn't. So, let's say that ad infinitum is just used to mean that the advancement will go on for a long time!

THE UNIVERSE ISN'T JUST A BUNCH OF ROCKS

A talk was given at a technical symposium that I attended around 1980 where the presenter stated a principle that I think is about the closest to the truth about future improvements of computer capability. The speaker observed that a particular storage media capability which had great expectations had failed. This was an approach called "bubble memory." It turned out that bubble memory didn't work out. This caused no problem in the progression of storage capabilities since another approach immediately took over. He then stated that:

"There will be no limit to the future of computer capabilities since, when a limit of capability of one architecture is reached - another one will always be discovered to take its place."

I have watched the progression of capabilities and have noted that, since then, this has certainly, consistently proved to be true. There have been many predictions of doom that have, so far, not proved to be true.

Whenever a stumbling block in technology has been reached, new technologies have been discovered to keep the trend going.

As the capabilities of computers, servers and the Internet increase, there is almost a feeling that they can do anything - and some day they probably will, "almost."

Computers are in most appliances, cars, and all devices. They are starting to communicate with each other. In the near future, we will probably have "Virtual Reality," where people are sort of *magically* transported to the city of their dreams (see later discussions).

A philosophical look at the advance of computer capabilities

Use the philosophy of Phydefiteration, making a prediction for the future based on analysis of past performance.

THE WILDLY FANTASTIC FUTURE BECAUSE OF COMPUTERS AND ADVANCED TECNOLOGY

You are led to the conclusion that computer capabilities of all kinds will continue to double at a rate of at least every 2 years and, most likely, at least every 1¼ to 1½ years - for many, many years to come.

What about the future?

So, let's see what we are talking about:

Everything doubles each 15 to 18 months, let's say 18 months.

In 15 years, this is an increase of 1000 (don't challenge me, experts, by saying 1024). In 30 years, this is 1,000,000. In 45 years, this is 1,000,000,000.

Just think of what a computer will be able to do! It will seem like

magic and will be able to do things like magic - but it's not magic, it's technology. People will also live for hundreds of years so they will be able to enjoy them for a lot longer (see later discussion).

A look at "magic" in the future

To help put things in perspective, let's revisit the folk story of the adventure tale of "Ali Baba and the Forty Thieves" from *One Thousand and One Nights.*

In this story, Ali Baba discovers the 40 thieves' treasure which is stored in a cave, the opening of which is sealed by **magic**. It opens on the Arabic words, *Iftah Ya Simsim - Open Sesame.*

This delightful story has been viewed as pure fantasy with magic controlling the opening to the cave. However, considering today's advances in technology, it now isn't very difficult to build a powered door which opens with a voice command such as *Open Sesame*. In fact, hardly anyone would give it a second thought if they saw it. (Store doors commonly open as we approach and voice commands are used more and more as everyday tools.)

THE UNIVERSE ISN'T JUST A BUNCH OF ROCKS

In the future, many, many things will happen, appearing to be **absolute magic** when they are, in fact, done easily with the available technology.

Some near-term future computer capabilities

Computers already are beginning to look different than the old concept of a desktop computer remaining in a fixed location. Laptops, tablets, and Internet-capable phones give you a basic idea of the future portability of what we call computers.

In addition, many household and portable objects will have access to the computers that are needed to do some of the advanced computer functions. It will be possible to go to the *refrigerator* or to the *front door* and talk to your computer. Your *car* will have a seamless interface to the computer system - even while turned off. Your sink, table or even the toilet should be seamlessly connected via networking.

One word of advice, when talking to your toilet, you might want to "keep it down" since there may be some "non-technical" people who may not completely understand when they hear you telling the toilet computer things like *control*, *alt*, *delete* or, even *reboot*.

Now, how do you control your computer and obtain the functions that you need?

It should be fairly straightforward to provide complicated voice commands. An easy way to envision this is to think of how you now access the Internet for search functions.

You simply type in a question and Google or some other search engine gives you your answer, perhaps with a couple of other selections.

The future common use of voice communications seems certain although fledgling attempts such as automobile voice-command control and Apple's Siri still leave much to be desired in terms of

THE WILDLY FANTASTIC FUTURE BECAUSE OF COMPUTERS AND ADVANCED TECNOLOGY

providing real, "fully understandable two-way communications."

Also, think of this as using much more advanced querying and control functions. These are very straightforward functions to develop.

You tell the computer what you are planning to do today, what your schedule is and add several caveats such as "make sure to warn me if I start to say something insulting to my boss (or my wife)." You may also say "adjust my schedule if xxx happens" or "tell me where the best looking girl is (or guy, as the case may be), in the area."

Now, you also have set up general rules for the computer to follow so you don't have to keep telling it with each situation. Pretty soon, you don't have to tell it much in the normal course of the day.

One of the best things that I can think about when considering all of this is that I can build an understanding with the computer. It can keep track of my past decisions and come up with a lot of suggestions for me in any new situation. Computer programs already do the rudiments of this in word-processing programs. Each of your devices, including appliances should be able to learn and apply the results of these "lessons" to future situations. Your lawn sprinkler system certainly should be able to tell when the lawn needs to be watered instead of you setting it to a fixed schedule. In short, most things should be automatic. You have already seen many examples of this. For one, your microwave knows how to defrost - it didn't a number of years ago.

There are, of course, all kinds of artificial intelligence functions that the computer can easily perform.

Think of some of the future things that computers will be able to do. I can't imagine that many people will ever need to sit down at the "den computer" for long periods in the near future.

The general rule for forecasting what the computer will be able to

THE UNIVERSE ISN'T JUST A BUNCH OF ROCKS

do for you in the future is this:

However much you can imagine the computer doing in the future, multiply that by 100 or 1000 or more.

Medical science future developments

Medical science has advanced at a rate approaching that of computers.

The key to understanding the extent of advances is to understand the past and extrapolate to the future.

The major breakthroughs needed include the following things that should be attainable in the not too distant future.

1. Disease cures, including cancer and all of the various diseases limiting human life.

2. Body repair.

    The use of automated repair functions is rapidly increasing. I see the future where little robots routinely travel throughout the body and diagnose and actually repair the defects found. Dispensing of drugs should be ideally suited for the computers controlling these functions to automatically keep levels optimum. It is easy to see, for example, how the applying of artery repairs for blocked arteries could be handled by these devices. Certainly, the suturing (probably using glue) of torn or open arteries or organs should be "duck soup" for these little guys.

3. Organ and system replacement

    Artificial organs should take a few years to become routine but there is no reason that this won't occur.

4. Solve the aging limitation

THE WILDLY FANTASTIC FUTURE BECAUSE OF COMPUTERS AND ADVANCED TECNOLOGY

Life span is currently something like a usual limit of about 114 years of age. The oldest person in the world is maybe 120 years old. The basic problem seems to be the aging process limits that are sort of built into our cell and tissue structure. While life expectancy has been rapidly increasing over recent decades, the life span has remained almost constant for thousands of years. Many potential breakthroughs seem to be on the horizon which will allow this to be increased.

(See a later section on this subject.)

Some more distant future computer capabilities

The key to understanding what these technological breakthroughs will be is to understand the past and extrapolate to the future.

In the past 50 years, several significant advances have been made but many of the advances have been in electronics, computers and the Internet. Transportation has significantly increased in speed but mostly through applications of previous technologies. A significant development has been the application of satellites and the accompanying GPS technology.

Significant new developments should be expected in the next few years.

In the distant future, computers will make everybody (well, practically everybody) into a Merlin, the magician.

You will just mention something that you want done and your computer will do it for you.

THE UNIVERSE ISN'T JUST A BUNCH OF ROCKS

Now, to try to get some perspective. Think about how people a few years ago would have felt if they had heard you say this: "Find me the best route to drive to see my friends in St Louis, MO." Almost instantly, the computer would have shown, in color, the best route with all way points, miles between each stop and times, etc. Then, "Take me there."

When they saw the GPS system in your car tell you, out loud, how to make each turn and where the McDonald's were along the way, they would have bowed down and worshipped you as a Greek God, at least.

There are many other examples, so try to imagine some in your area of interest.

In the future, you will be able to say "get me some food" or "show me my grandchild doing yoga" or "find the only person in the world with my matching, exact blood type and tell me all about him/her."

It will be very simple, technically, to do remarkable things, for example to make things invisible. All that you would need is to have camera-type devices photographing things around you. Then, the scenes on each side of you could be displayed as though you and your vehicle weren't there. Every pixel of light- illuminated scenes around you could be transposed to the other side to whatever viewers were there. In the same way, other kinds of electromagnetic waves could be similarly handled so as to make you invisible to anyone else. Even anyone's radar and other equipment could be completely fooled. The secret is that computers would have the capability to process each tiny bit of information, such as pixels of light or "grains of sand" and handle them individually and complexly.

THE WILDLY FANTASTIC FUTURE BECAUSE OF COMPUTERS AND ADVANCED TECNOLOGY

It is easy to imagine computers making many factories unnecessary. Why put together shoes in China and ship them to the U.S., then place them in a store staffed with sales clerks where you travel to the mall and try them on?

To start your thinking, think of ordering over the Internet.

In the future, why not just tell the computer what you want. Have it provide you with a virtual set of shoes which you can try on - virtually and select a pair. Then the computer can not only order the shoes but tell the automated factory to build the pair for you and automatically ship them to you via automated facilities. You should get the ones that you want overnight or less at an inexpensive price.

Of course, in the distant future, you can have a general purpose machine in the garage or the basement which will create the raw materials and manufacture the shoes for you. This machine should be able to even build you another machine to make something else, if needed.

Basically, the world is your oyster but, a word of warning; this may take 40 or 50 years to realize (up to 100 or even 200?).

The really distant future

This will be harder, but try to imagine, say 200 to 500 years from now. Computers may be able to handle almost anything. Nobody in developed countries will need to work since all food will be synthesized by the computer, Warmth, cooling, clothes, and transportation, even sensory pleasures may be handled by your trusty computers.

Virtual Reality (see later discussions) is to the point where it isn't necessary to travel to see your friends or cultural sites. It certainly won't be necessary to go to Tahiti to sit on the beach with the sound of waves lapping at your feet. That will be so "old hat" it will be pitiful!

THE UNIVERSE ISN'T JUST A BUNCH OF ROCKS

Time travel to the past will be common place. Time travel to "one of the possible futures" will be standard. Do you want to see (simulated) and talk to one of your relatives who died years ago? No problem. Just mention it to the computer.

What are some possible stumbling blocks to "unlimited advances" in the future?

A real need for the distant future is to find possible new Universe *rules*, or *adaptations* or *alternative approaches*, such as:

1. Methods must be found to increase the speed of transmission of data, now limited by light-wave propagation speeds.

2. Means must be found of exceeding the speed of light and other waves. This, especially, is needed for future space travel - real or virtual.

3. New ways of increasing propulsion for vehicles. (Think of "Warp drives", etc.)

4. A possible method of circumventing gravitational restrictions.

5. New transportation possibilities such as flying cars or the equivalent.

6. Any other of the well-known science fiction breakthroughs would also be welcome. (However, a Star Trek "Transporter" is not visualized as one of the possible ones, at all! Of great concern is the problem of taking a part all of the molecules in the body and putting them back together. This also leaves concerns about what happens to the *soul* in this process.)

THE WILDLY FANTASTIC FUTURE BECAUSE OF COMPUTERS AND ADVANCED TECNOLOGY

While the future of computers, medical technology and related technologies are seen as certainly advancing in fairly known directions, the above-mentioned technological developments are not quite so certain.

Let's try to get the future prognostications into perspective

Now, let's be reasonable. The preceding specific examples of the future may partly come to realization, but we can be sure that not **all** will, in exactly this way. There will be others, which haven't been forecasted that will happen. Some things will be difficult and slower. Some things won't happen for a few years longer.

BUT, and please get this BUT; the general idea of fantastic future advances in many, many ways WILL happen. Also, it will happen in the fairly near future.

And I am not talking about 1000 years from now, either - for sure!

I do have to mention that they probably won't happen or be applied to all people, due to, what I will just call *the inequities of our civilization*.

One final comment is this: Think about the possibility of other humankind, on other planets in the Universe. If they exist, their development should also lead to the same fantastic developments in their futures. They will also have thousands of years for development after they reach a stage of computers, which I feel sure that they will come to. (See later descriptions of possible "aliens.")

THE UNIVERSE ISN'T JUST A BUNCH OF ROCKS

14 Getting There Faster Than at "Light Speed"

This subject is worthy of discussion separately from the previous predictions of computer, medical science, and other technology advancements. While details in all of these areas may not be too accurate, in general, things are seen as almost certain to occur.

The area of exceeding the speed of light for communications and for actual, physical travel is more speculative. Considerable theoretical study has shown some hope but no real breakthroughs have yet happened.

The first thing to make clear is that I am not really talking about exceeding the speed of light. What I am discussing is getting the capability to communicate or to physically travel from one point to another faster than an electromagnetic wave could do it.

This probably means "jumping" from point A to point B.

An author's biased prediction about exceeding the speed of light

I have tried to be fairly level- headed and reasonable with most of my guesses about the future.

I do have one personal biased prediction which I will allow myself to present to you. It isn't based much on knowledge or even reason. It is mostly based on an understanding of the need that we will have in the future for realizing many of our technical goals, including space exploration. I feel that the creator of the Universe had plans and that those plans included people as a key ingredient. I feel that people need to be able to understand things and gain knowledge and that the creator will help them.

THE UNIVERSE ISN'T JUST A BUNCH OF ROCKS

What is really necessary to understand these things is for people to be able to find a key to "Exceed the Speed of Light" in the Universe rules/laws. Doing this includes exceeding the speed of other waves, and perhaps the effective "transmission" times for gravity effects, and who knows what else.

The Universe was set up with all of the elements and the rules/laws being very integrated and coordinated. Everything was well thought out. That is, everything works together giving evidence of a complete plan. Nothing was left out. Just think about how well all of the pieces work together.

It doesn't seem logical that the Universe would have been allowed to continue to expand with no methods for anybody to be able to transverse the many light years that separate galaxies.

In other words, it seems reasonable that the plan - the Universe rules - would not have limited people's ability to travel or to communicate at simply the speed of light.

There simply have to be ways for people to communicate and to travel in ways that will span the huge numbers of light years both within and between galaxies.

So, scientists, start using outside-the-box methods, new concepts, seemingly crazy ideas, and genius approaches. Just do it!

Figure out how to **exceed the speed of light**. It can be done.

Possibilities for getting from one place to another faster than a light wave could

Again, it is stressed that these things are theoretical and speculative.

GETTING THERE FASTER THAN AT "LIGHT SPEED"

Wormholes

A wormhole is a **hypothetical** interconnection between two regions of space-time. It would be, fundamentally, a shortcut through space-time. For a simple visual explanation of a wormhole, consider space-time visualized as a two-dimensional surface. If this surface is folded along a third dimension, it allows one to picture a wormhole bridge. A wormhole is, in theory, much like a tunnel with two ends each in separate points in space-time.

In 1935, Albert Einstein and Nathan Rosen realized that general relativity allows the existence of bridges, originally called Einstein-Rosen bridges but now known as wormholes. These space-time tubes act as shortcuts connecting distant regions of space-time. By journeying through a wormhole, you could travel between the two regions faster than a beam of light would be able to if it moved through normal space-time. There is no observational evidence for wormholes, but on a theoretical level there are valid solutions to the equations of the theory of general relativity which contain wormholes.

Until recently, theorists believed that wormholes could exist for only an instant of time, and anyone trying to pass through would run into a singularity. But more recent calculations show that a truly advanced civilization might be able to make wormholes work. By using something physicists call "exotic matter," which has a negative energy, the civilization could prevent a wormhole from collapsing on itself. Perhaps, some day in the far future this could be realized.

Warp speed

The **Alcubierre drive**, also known as the **Alcubierre metric**, is a speculative, but valid solution of the Einstein field equations. It is a mathematical model of space-time exhibiting features reminiscent of the fictional warp-drive from *Star Trek*, which can travel faster than light, although not in a local sense.

THE UNIVERSE ISN'T JUST A BUNCH OF ROCKS

In 1994, the Mexican physicist Miguel Alcubierre proposed a method of stretching space in a wave which would in theory cause the fabric of space ahead of a spacecraft to contract and the space behind it to expand. The ship would ride this wave inside a region known as a *warp bubble* of flat space. Since the ship is not moving within this bubble, but carried along as the region itself moves, conventional relativistic effects such as time dilation do not apply in the way they would in the case of a ship moving at high velocity through flat space-time relative to other objects. Also, this method of travel does not actually involve moving faster than light in a local sense, since a light beam within the bubble would still always move faster than the ship; it is only faster than light in the sense that, thanks to the contraction of the space in front of it, the ship could reach its destination faster than a light beam restricted to travelling outside the warp bubble. Thus, the Alcubierre drive does not contradict the conventional claim that relativity forbids a slower-than-light object to accelerate to faster-than-light speeds. However, there are no known methods to create such a warp bubble in a region that does not already contain one, or to leave the bubble once inside it, so the Alcubierre drive remains a hypothetical concept at this time.

Superluminal communication

Superluminal communication is the term used to describe the hypothetical process by which one might send information at faster-than-light speeds.

On the other hand, what some physicists refer to as apparent or effective faster-than-light speeds is the hypothesis that unusually distorted regions of space-time might permit matter to reach distant locations faster than it would take light in the normal or undistorted space-time. Although, according to current theories, matter is still required to travel subluminally with respect to the locally distorted space-time region, apparent faster-than-light speeds are not excluded by general relativity.

GETTING THERE FASTER THAN AT "LIGHT SPEED"

Examples of faster-than-light speeds proposals are: changing the frequency of mass to a higher state by applying high frequency waves of energy, the Alcubierre drive, and the traversable wormhole, although the physical plausibility of some of these solutions is uncertain.

Other:

There are almost certainly many other possibilities....................

THE UNIVERSE ISN'T JUST A BUNCH OF ROCKS

15 The Virtual World of the Future

First, it is important to reiterate that this description of a Virtual World is very different from similar discussions in science-fiction novels.

These scientific stories, while sounding realistic are just that - stories or novels.

What is described here is real. Not only is this real and achievable but it is almost certain to be accomplished in the very near future.

Now, when anyone is predicting future events, some of the details of these are only guesses. However, the general ideas are fully expected to be highly accurate.

This estimate of accuracy of predictions is based on past realizations.

For example, I couldn't have predicted, a few years ago, the full impact and specifics of the provisions of the Internet.

But, I could then see the huge future potential of advanced computers, networks, and the input from many, many thousands of computer programmers and development people. I could also see that future advances are usually (as they have been) much greater and faster than expected.

To reiterate, advances in technology are usually many times greater than anticipated!

THE UNIVERSE ISN'T JUST A BUNCH OF ROCKS

What is a virtual world?

I will suggest a very sophisticated, but very accomplishable **"virtual reality"** system. Using this system, selected participants will be connected to a virtual world in the safety of a secluded vault. They will use computer technology which is not yet available but which is right around the corner - see later description of a reasonable timetable.

The virtual world is a computer-generated entity exactly (or almost exactly) copying a real-world location. The participant or participants will select a location of their interest - maybe 2,000 miles remote or perhaps many times farther away. All aspects of this selected location will be presented to the participants in visual three-dimensional forms together with sound, smell, vibrations and nerve feelings and responses. To the participant, he will seem to **be** at a spot in this recreated location. The virtual world will be created by the computer from satellite views, from remote pickups, and from other, yet to be developed, methods of viewing and sensing such locations. Sound will be captured. Smells and vibrations will be created by the computer based on information captured by these sensors and viewers. Sensations of smell and feel will be transmitted to the human nervous system directly so as to give the participant a feeling of reality.

If this seems farfetched, remember that we are talking about a few years from now with all of the advances which we know are coming in these fields.

The key design factor in evolutionary development of this Virtual World

The key factor in the creation and maintenance of this virtual world is the computer's ability to what we will call **"fill in the gaps."** In the scientific world this would be called interpolating, extrapolating or smoothing the virtual environment. This leads us to the understanding of the evolutionary method of developing

THE VIRTUAL WORLD OF THE FUTURE

virtual environments. In the earliest stages of development, the virtual environment will be based on as much information as is available. The computer will fill in the rest. It might be compared to virtual environments which are now presented in very simple form in video games, except that all images will be thousands of times more real. They will be presented as holographic, three-dimensional views. As the science of maintenance of virtual environments improves, more and more frequent updating of the computer-stored virtual environment will be done - until it is essentially a "real time" simulation of the real world. Feedback to the remote real environment can then take place. Outputs from the participants in terms of sound, movement, etc. will be actually presented to persons and "sensors" in the remote, real environment.

This kind of evolutionary development will be even more noticeable when creating and maintaining remote environments thousands and million miles distant - such as on the Earth's moon or her planets.

In understanding the evolution to this real-time interaction of the participants with the real-world's environment, speed of transmission of information will be key. This involves more than just data transmission limitations but also involves problems of distances involving the speed of light or radio wave transmission limitations from the virtual environment to the real environment.

See later discussions of ways to overcome these problems.

THE UNIVERSE ISN'T JUST A BUNCH OF ROCKS

How will the participants use the "virtual world?"

All of the senses of the participants will be connected, to the computer's inputs and outputs. The extent of this will be to shut out their normal senses. Connections will be made directly to the nerves representing these senses - through noninvasive connections. The optic nerve will be receiving all pictures, hearing, taste and smell will be direct. Feeling and skin and bone sensations will be via the computer output. Movements of the virtual (projected) participants will be controlled by the actual participants as their nerve signals are sent, and intercepted by the computer's inputs, to the limbs and other parts of the body.

The participant and other selected objects will be virtually transported to a virtual re-creation of another, perhaps distant selected location. To the participant, he will **be** in this other, location. He will **see** all people and objects around him. He will be able to virtually touch, feel and smell these things. Sound from the selected location will be heard by the participant. This sound will, of course, be handled by the computer and will be translated to whatever language the participant desires. As described before, when updating of the virtual environment from the real world environment is sufficiently fast, holographic images of the virtual participants and selected objects will be projected in this selected location using methods yet to be developed, but already conceptualized. For example, satellite remote transmission of these holographic images or using remote transmission facilities may be used. In the same way, sound from the participant will be broadcast to the other location, and, of course translated to the appropriate language(s) by the computer.

THE VIRTUAL WORLD OF THE FUTURE

What will force the development of this virtual world?

What are the forcing factors in causing the pieces of the Virtual World to be actually developed?

First, of course are video games. The corporate efforts in this area should cause many of the pieces of the Virtual World to come into being.

Next is the entertainment world. People are fascinated with movies and television and have shown their willingness to spend in these areas to almost an unlimited extent. The desires of millions of people should help force much of the needed development in the "Virtual World" direction.

Well, we'll see....................

Later add-ons

The effects of the movements of participants and the effects of the other computer outputs which are commanded by the participants will be a later addition to the system and will be accomplished by remote objects under the control of the computer. At the beginning of this development, participants won't be able to control remote objects but will be able to control *virtual objects*. *Virtual objects* are those that are generated and controlled by the computer, but which are not seen by the people at the actual location nor do they affect the actual environment.

Another future *possibility* for the virtual world might be the development of the capability to control human dreams through a computer interface, with the individual controlling the computer. Think of the power of harnessing the capability of the human brain to generate dreams which are controlled by the individual himself. The possibilities are almost endless and without needing any of the

THE UNIVERSE ISN'T JUST A BUNCH OF ROCKS

hardware which would be necessary for the virtual world. Your fantasies could be realized without any external repercussions!

What is needed in order to accomplish this?

1. Input interfaces to the participants, including holographic "rooms" which allow the participants to move within them. Input devices need to directly interface through the nerves to the brain rather than through the eyes, ears, touch sensors, etc. We are only beginning to approach such interfaces and there is a long way to go before these will be able to adequately accept inputs. "Output interfaces from the participants - to the images and to the computer. Control of virtual movements, et al. need to be through normal thought methods rather than through artificial means such as a mouse or speech. We are somewhat closer to some of these methods but still a way to go.

2. Projection of holographic images of the virtual participants and selected objects. Only in recent times has this kind of projections begun to become closer to feasibility due to advances in computer power and giant memory advances.

3. Broadcast of sound and vibrations from the virtual participants and selected objects.

4. Extremely powerful computer capabilities of all the types described above.

5. Extremely fast communications capability in terms of data rates. Also, programmed methods of simulating even higher data rates through prediction and other algorithms.

6. We need to be able to exceed the *speed of light*. No problem, you say? It's quite a problem but we will do it! Of course, simulated methods of exceeding the speed of light will help until the day when, perhaps, real methods are found.

THE VIRTUAL WORLD OF THE FUTURE

Let's see an example

The participants are hooked up to the system in a secure, environmentally-controlled vault. They start by using computer selections, to select themselves and other desired objects for their "basic environment." This basic environment would consist of the persons, clothes, and any other items to accompany the participants. For example, a car might accompany the participants.

Through computer selections, they then ask for this basic environment to be "virtually transported" to Paris to near the base of the Eiffel Tower at the corner of the International Hyatt Hotel. This means that this virtual environment will be created by the computer, presented (displayed) to the participants and updated to continuously represent the selected, remote real location. The images of the actual location are projected, in this virtual environment, to the participants in full three-dimensional form together with sounds, smells, etc. They might start by using computer commands to move to a wine shop across the street. After examining a number of wines and discussing wines with the proprietor, they would ask the computer to provide several virtual bottles of selected wines. They could store all of these in a thimble-sized pouch on their belt. The computer will also keep track of things such as this in order to be able to automatically compensate the wine shop proprietor for his time, loss of sales and to the region or country for duties, etc. (We will worry about this kind of thing at some later time.)

Next, the participants might like to travel to the top of the Eiffel Tower. Why use the elevator? They will just rise slowly or very fast, to the top. Perhaps they will even get several hundred feet above the Tower in order to better view it and maybe snap some pictures. They won't be bothered by wind, if they don't want to - just ask the computer to filter it out. If the day is cloudy and overcast - no problem, the computer can make it a pretty day.

THE UNIVERSE ISN'T JUST A BUNCH OF ROCKS

They use the computer to hail a *virtual taxi*. They then instruct the *virtual driver* to take them to a particular restaurant where they occupy a *virtual table*. If the space of the *virtual table* is actually occupied by other actual people or actual objects the computer will, of course, move the *virtual table* or obscure the actual objects. A *virtual glass of wine* would be fine - let's open one or two of our bottles! The computer system will always select the exact bottle for the participant, perhaps, even improving on the taste according to the desires of the participant.

Let's even suppose that the participants don't care for the restaurant or, even for Paris, today. They could ask to be relocated to Tahiti, which would take, maybe because of *rush hour traffic*, several seconds.

As previously mentioned, virtual **time travel to the past** will take advantage of the knowledge of history PLUS the interpolation and extrapolation of information by advanced computer applications. Libraries of books and written materials, placed online, are scoured by the computer applications. Complete stories together with videos are made available to the users through the virtual world.

Virtual **time travel to the future** will take advantage of history but will use the computer prediction applications to make educated and highly accurate predictions of future "Histories."

The future applications may take longer to develop than the history compilations. However, I am talking about *maybe 50 to 75* years, not *500*.

Exceeding the speed of light limitation

The problem with exceeding the speed-of-light limitations will be key to making Virtual Reality of real situations really work.

THE VIRTUAL WORLD OF THE FUTURE

Let's say that when we were located in San Francisco, California, we wanted to visualize actions at a bistro in Paris - using cameras, image projections, and long distance communications. Then we wanted to react to these actions and perform some actions in response - in near real-time.

Or, perhaps, we wanted to remotely control a robot on the moon in real time, without this "speed of light" delay.

In the Paris example, we might use communication satellites. These may be at an altitude of 20,000 miles. This means that there is a delay in transmitting even one bit of data of over 1/5 second from one point on the Earth to another point (20,000 miles /186,000 miles per second of the speed of light, to the satellite, and then back to the Earth). Of course, distances to the moon or to planets are much greater. Average distance to the moon is about 238,000 miles. It takes a radio wave 1.28 seconds to travel to the moon. And, to get there and back, it takes about 2.5 seconds. Of course, traveling to more distant planets or to other distant points may involve minutes, hours, days, even years.

How could we solve this limitation?

The scientist would say that it is impossible.

Rex Stout's, Nero Wolfe would say, "Pfui!"

The philosopher would say, "How about another approach?"

There are many possible solutions. The most obvious is to simply use prediction from one transmission to the next. For example, a car is moving down the street at a speed of 30 miles per hour. There is a delay of 2.5 seconds for starting a satellite transmission and 1 second to transmit the next image. The computer could easily calculate the new image of the car and its location 3.5 seconds from now and not need the new image to be translated.

THE UNIVERSE ISN'T JUST A BUNCH OF ROCKS

Predictions of the beautiful Parisian girl at the bistro might show her desiring a glass of red wine. Knowing her predilections, I might anticipate her desires and order the glass of wine before she asks. The waiter responds to our order and delivers the wine immediately as the request comes from her (beautiful) lips. She smiles sweetly at my image in the chair next to hers and I, anticipating her action, knowingly smile back. She jots her address and phone number on a slip......oops, I got carried away. That's enough of this example for now. I've got to get to the airport - fast.

The robot on the moon can be controlled in a like manner. We want to command the robot to pick up a rock of a certain size. As the robot starts to reach for the rock, our next command is anticipating his movement two or three seconds from now. We also see which rock is of the right size and get him started. As he reaches (for the wrong rock), we know where he is heading and command him to change to another rock before he gets there. We need to stay ahead of each action of his just enough to account for the time delay.

These are very rough examples of one approach which might be taken to overcome the limitations of the speed of light. While there are certainly limitations to what we can achieve, there are fantastic advanced possibilities when considering development of computer artificial intelligence and the huge advances in computer speed and capabilities themselves.

THE VIRTUAL WORLD OF THE FUTURE

Evolution of the System

The first versions of the system will utilize simplified versions of the above-mentioned features - especially of the interceptions of nerve inputs and outputs from the participants. There is a later description of time tables in this book. These time tables appear very reasonable and assume an approximate starting date for a beginning version of this system, using a recorded virtual environment and very basic visual and sound interfaces. The computer would fill in many of the gaps and there would be no real time update.

In part, this example might be realized about **2022 AD**. A very good version of the system as described above might be capable of being created about **2040 AD**.

Doesn't it sound exciting?

Then, what's next?

Of course, there is the possibility of a Virtual World where it is totally not real. The individual is actually part of the action, in this case. This is, of course, the extension of the Game Worlds of videogame extractions.

Then, there are a myriad of possibilities of "tailored" worlds. These might start with a real environment or even attempt to closely follow the real world.

A natural extension is the Virtual Time Machine - taking the user into the past, but making it more realistic using all of the information available to the computer from thousands of documents. (Imagine, for example, using all of the information available on the Internet 20 or 100 years from now.)

Then, there is the Time Machine foraging into the *future*. There are almost limitless possibilities, but imagine the future possibilities perhaps 100 years from now!

THE UNIVERSE ISN'T JUST A BUNCH OF ROCKS

Virtual dreams

Another, future *possibility* for the virtual world might be the development of the capability to control human dreams through a computer interface, with the individual controlling the computer. Think of the power of harnessing the capability of the human brain to generate dreams which are controlled by the individual himself. The possibilities are almost endless and without needing any of the hardware which would be necessary for the virtual world. Your fantasies could be realized without any external repercussions!

If you want to get fanciful, think of a "split screen" type of dream, with one screen controlling the dream and the other screen realizing the dream. Put in a "pause" function of the realizing part while the control part adjusts things, and you're all set. (You'd never want to go to the movies again.)

It might also be prudent to establish an "editing" function. Using this capability, a person could erase those dream sessions that didn't work out too well. He or she could store, in their entirety, for easy recollection - in sort of "bold type," those that he or she wanted to remember. Of course, the replay capability would be great too.

16 Our Space Travel in the Universe

For many people space tourism and even colonization are attractive ideas. But in order for these to start we need vehicles that will take us to orbit and beyond and bring us back. Current space vehicles clearly cannot. Only the Space Shuttle survives past one use.

There is an indication that private companies may be more involved in space exploration in the future.

An example is the company, SpaceX. Plans are to try to provide reusable vehicles to become available for passengers in the time frame of about 2014-2015. China is working on planned developments also.

Space travel is not an easy endeavor as we all know. To get into orbit requires accelerating to Mach 26, and so it uses a lot of propellant - about 10 tons per passenger. This means that the cost per passenger will remain quite high for the foreseeable future. Safety is also a major problem requiring significant development time.

So let's make an effort to estimate time frames for space travel, given the present technology.

The dream of routine flights for a weekend destination seems destined to wait for significant improvements in technology. For example, this seems to necessitate reducing the need for 10 tons of propellant per passenger quite significantly.

We also need to remember that we have only taken people barely out of orbit around Earth to one location. These trips to the moon (a close body, by the way) were very risky.

There are many safety and associated problems with "out of Earth orbits," including radiation, to say the least.

THE UNIVERSE ISN'T JUST A BUNCH OF ROCKS

There doesn't appear to be sufficient present technology, at present, so this rapid growth of flight for the common man seems to be in the future. This, somewhat pessimistic picture is quite markedly different from the forecasted huge advances in Virtual Reality, medicine and other computer-related technology.

The future time line for space travel

Perhaps the following might be somewhat reasonable:

We might hope for, say, a cost of $20-million per passenger for travel to other close bodies in space.

We might see a colony on Mars in 25 years, but with a huge cost associated with it.

Perhaps we might make a flight to another solar system in 40 to 50 years, but not for everybody.

With the use of robotics, we might envision much faster "virtual" space travel in a much quicker time frame. Remember, virtual space travel requires only one fixed cost for the vehicle and the flight plan with many, many people then able to "use the facilities" without multiplying the cost proportionally per "passenger."

In addition, with robotics, we eliminate many problems and simplify the requirements for super safety.

What is needed for real space travel?

For real space travel to truly come of age, we need some significant breakthroughs.

1. We need significant methods of cost reduction, especially in the amount and cost of propulsion.
2. We need acceleration without the huge propulsion needed at present (slightly different from above).

3. We need acceleration methods without the huge forces on humans.

And,

4. We need to be able to exceed the speed of light for future, other solar-system excursions.

Once we can exceed the speed of light, the Universe is our oyster!

> *Probably, the way of exceeding the speed of light limitation would mean determining how to jump across regions of space; for example, using hypothetical "wormholes" or some other method of simply travelling from one region of space to another. Whatever method could be used, it must be something different from travelling by normally propelled rocket ships.*

Great advances don't seem too probable, for the very near future.

Perhaps in 50-to-100 years?

THE UNIVERSE ISN'T JUST A BUNCH OF ROCKS

17 Can We Contact Other People in the Universe?

People have spent a great deal of effort and lots of dollars on this subject.

There are really two questions: First, are there other human-like forms in the Universe and secondly, can we contact them or can they contact us?

Is there other life in the Universe?

When considering the vastness of the Universe, it is difficult to consider that ours is the only planet that has life. Statistics showing the number of inhabitable planets alone would suggest that there may be others out there.

Also, if a planner and creator created the Universe and provided for the creation of people as one of his principle objectives, would he have created such a vast Universe for just the people on Earth? Talk about getting too little *Big Bang for Your Buck*, wow!

What planets are habitable?

What do planets need to be able to support life?

Just to get an idea, we can extrapolate some of the requirements of Earth. Some necessary features would be:

THE UNIVERSE ISN'T JUST A BUNCH OF ROCKS

A solar system is needed, with a single sun than can serve as a long-lived, stable source of energy. A solid, nongaseous planet is needed. It must be close enough to the sun that water can be preserved at the surface and the water is in liquid form - not frozen. It can't be too close or the water will evaporate. The solar system must be placed at the right place in the galaxy - not too near dangerous radiation, but close enough to other stars to be able to absorb heavy elements after those neighboring stars die. It needs a moon of sufficient mass to stabilize the tilt of the planet's rotation. It needs an oxygen-rich atmosphere without poisonous concentrations. There needs to be associated planets, with non-eccentric orbits that can deflect comets.

The idea that planets beyond Earth might host life is an ancient one, though historically it was framed by philosophy as much as physical science. The late 20th century saw two breakthroughs in the field. The observation and robotic spacecraft exploration of other planets and moons within the solar system has provided critical information on defining habitability criteria and allowed for substantial geophysical comparisons between the Earth and other bodies. The discovery of extra solar planets, beginning in the early 1990s and accelerating thereafter, has provided further information for the study of possible extraterrestrial life. These findings confirm that the Sun is not unique among stars in hosting planets and expands the habitability research horizon beyond our own solar system.

How many planets are there?

There are estimated to be 200- to- 400-billion galaxies, with many hundreds of billions of stars in each galaxy. Let's estimate that one in a million stars would be classed as planets. That's something like ten-million-billion planets. Then, let's guess that one-in-a-million planets would be capable of supporting life. This would give us perhaps ten-billion planets capable of supporting life.

CAN WE CONTACT OTHER PEOPLE IN THE UNIVERSE?

The actual numbers aren't too important and are only wild guesses anyway.

However, the point is that there is a whale of a lot of planets capable of supporting life. Even if the number is only one billion, that would seem to be enough for life to be possible.

There are two possible explanations for how life formed on Earth

1. Life formed purely by chance.

2. Life was created by the creator of the Universe, who I call God.

If life started by chance on Earth, there is an excellent probability that it formed on other planets. The conditions were similar and, at least in many cases, extremely similar to those on Earth.

Now, if life was created by God on Earth, there must be a very strong suspicion that he created it elsewhere also.

Why would he create this vast Universe with billions and billions and on and on of everything and only create one Earth with people on it?

As stated before, it seems like the greatest of overkill that I can possibly imagine.

Is it just so that we can sit on a hilltop on a summer eve, with our boy/girlfriend, and star gaze at the heavens?

Now, I know that the bible said that God created us, in the story of seven days, with Adam, Eve and that nasty old snake. However, the biblical account also stated that he created us in his image, when he didn't have one - since he was in the non-universe where there was no matter *before* the Universe was created. The bible story also says that "on the first day" he did such and such. Well, on that first "day" there weren't any "days."

Then there were two lights - one to light the day and the other to light the night.

It seems obvious that the biblical story was written 2000 years ago, or so, by people telling a simplified account of the actual creation. After all, they didn't mention anything about evolution or the "Big Bang" or the "time dilation due to gravity or velocity" or a host of other very germane things. It seems most probable that the creation of people on Earth was done through a long evolutionary period.

It does seem that there is a good chance of there being other people, or, at least other "life forms" in the Universe. This would be true, either if life formed by chance or if God created man (and, I hasten to add, women).

That's my story and I'm sticking to it!

Can we contact these other "beings" in the Universe?

The search for extraterrestrial intelligence (SETI) is the collective name for a number of activities people undertake to search for intelligent extraterrestrial life. Some of the most well-known projects are run by the SETI Institute. SETI projects use scientific methods to search for intelligent life on other planets. For example, electromagnetic radiation is monitored for signs of transmissions from civilizations in other worlds. The United States government contributed to early SETI projects, but recent work has been primarily funded by private sources.

In November 1961, ten radio technicians, astronomers, and biologists convened for two days at Green Bank. Young Carl Sagan was there, as was Berkeley chemist Melvin Calvin, who received news during the meeting that he had won the Nobel Prize in chemistry.

It was in preparing for this meeting that Frank Drake came up with what soon became known as the Drake Equation:

CAN WE CONTACT OTHER PEOPLE IN THE UNIVERSE?

$$N = R \times f_p \times n_e \times f_l \times f_i \times f_c \times L$$

It expresses the number N of "observable civilizations" that currently exist in our Milky Way galaxy as a simple multiplication of several, more approachable unknowns:

R is the rate at which stars have been born in the Milky Way per year; f_p is the fraction of these stars that have solar systems of planets; n_e is the average number of "Earthlike" planets (potentially suitable for life) in the typical solar system; f_l is the fraction of those planets on which life actually forms; f_i is the fraction of life-bearing planets where intelligence evolves; f_c is the fraction of intelligent species that produce interstellar radio communications; and L is the average lifetime of a communicating civilization in years.

Astronomers and biologists alike have tried to "solve" the equation ever since.

These and other efforts have been launched to study whether other "people" might be sending radio signals or signaling in some other ways. Even now, there are many projects looking for signals from outer space, signals which would indicate the existence of other humanoid creatures.

The "Flying Saucer" craze and UFOs

The "Flying Saucer" craze began with an incident on June 24, 1947, when private pilot Kenneth Arnold said that he spotted a string of nine, shiny, mostly disc-like unidentified flying objects flying past Mount Rainier at then unheard of supersonic speeds that Arnold clocked at a minimum of 1,200 miles an hour.

THE UNIVERSE ISN'T JUST A BUNCH OF ROCKS

This resulted in the creation of the term, flying saucer, by U.S. newspapers. Although Arnold never specifically used the term, flying saucer, he was quoted at the time saying the shape of the objects he saw was like a saucer, disc, or pie plate, and several years later, added that he had also said "the objects moved like saucers skipping across the water." Both the terms, flying saucer and flying disc, were used commonly and interchangeably in the media until the early 1950s.

Arnold's sighting was followed by thousands of similar sightings across the world. Such sightings were once very common, to such an extent that "flying saucer" became a synonym for UFO or Unidentified Flying Object.

More recently, the flying saucer has been largely supplanted by other alleged UFO-related vehicles, such as the black triangle.

The Roswell UFO Incident was the recovery of an object that crashed in the vicinity of Roswell, New Mexico in about June of 1947. It was alleged to have been an extra-terrestrial spacecraft with alien occupants. Since the late 1970s the incident has been the subject of intense controversy and of conspiracy theories as to the true nature of the object that crashed. The United States Armed Forces maintains that what was recovered was debris from an experimental high-altitude surveillance balloon belonging to a classified program named "Mogul."

The term UFO was invented so as to include a wide variety of shapes being reported to have been observed. However, unknown saucer-like objects are still reported, such as in the widely-publicized 2006 sighting over Chicago-O'Hare airport.

(There were also stories about the Germans building some flying saucer prototypes during World War II, using standard propulsion means.)

CAN WE CONTACT OTHER PEOPLE IN THE UNIVERSE?

Many of the alleged flying-saucer photographs of the era are now believed to be hoaxes. The flying saucer is now considered largely an icon of the 1950s and of grade B movies and comic science fiction.

The question is, "should we believe that we are apt to contact others from other planets?"

Contacting "Extraterrestrial People" - let's think about this.

There are two problems. First, consider the limitation of the speed of light.

When we say that we are looking for radio signals or other signs of interplanetary life, we really mean that we are looking for signs of where they may have been a long time ago. The nearest star to our Sun is four light-years away. The nearest galaxy to ours is the Canis Major Dwarf galaxy, about 25,000 light-years away.

When we are looking for signals from them, we are looking for signals from four years ago or, more likely from *25,000 years ago to billions of years ago*.

If we found that there was life on another habitable planet, we would find that our discovery was out of date by thousands or, more likely, millions of years.

So we wouldn't be actually contacting these people at all. It would still be interesting and have a number of possible ramifications in our lives, but we aren't going to answer the radio broadcasted messages using cell phones at each end.

Then, we are sending radio signals into space. Who do we think is going to receive them?

THE UNIVERSE ISN'T JUST A BUNCH OF ROCKS

It will certainly be someone a long time into the future because of the speed of radio-wave transmission delays and the fact that anyone who might be able to receive signals is many light years away.

In addition to the "speed of light" problem there is another problem with the idea of contacting other human kinds in the Universe. Many of the currently attempted methods of contacting other people in the Universe demonstrate a BASIC mistake in judgment because of the relative "Time Lines" of evolutionary development of people on Earth and on other "planets."

Time Lines of development of possible extraterrestrial beings

Let's take a look at the question of the other "people" (or beings) that might be out there somewhere.

First, look at the basic time lines of beings coming into existence.

The Universe development used evolution of life forms of development to make things happen so it's reasonable to assume that the "creator" would have done this in creating multiple races of people on different planets. The creator would have used evolutionary development so that the time lines of development would have roughly followed the schedules of planetary development. The development of life would have followed, roughly, the planetary schedule - it is assumed.

In the same way, *chance* development of life would have certainly followed the solar system and planetary development schedules.

In general, I will assume that the evolutionary process evident in the formation of the Universe, including our galaxy and our solar system would be followed with similar steps throughout the Universe.

CAN WE CONTACT OTHER PEOPLE IN THE UNIVERSE?

Different galaxies, solar systems and, planets were created at different times, following the expansion of the Universe. Following the apparent scheme of the "Big Bang Theory" (or whatever theory you like), other planets or bodies supporting human life development would have then seen this development according to somewhat similar, but probably different schedules and rates - but at a different starting time than the one on Earth.

It would be reasonable to assume that the basic evolutionary rate of progress would follow similar steps in the development of each human-like race.

What is the age of the Universe?

The age of the **universe** is estimated to be about **14-billion years**.

The age of our **solar system** is estimated to be about **4.6-billion years** when it was formed from the collapse of a giant molecular cloud.

How old is man?

The Earth is estimated to be about 4.5-billion years old. Maybe, three-billion years of photosynthesis, one-billion years of multi-cellular life and 600-million years of simple animals.

Then 475-million years of land plants, 300-million years of reptiles, 200-million years of mammals, 150-million years of birds, 130-million years of flowers.

Perhaps 65-million years since the non-avian dinosaurs died out.

Then 2.5-million years since the appearance of the genus Homo.

200,000 years since humans started looking like they do today and 25,000 years since Neanderthals died out.

THE UNIVERSE ISN'T JUST A BUNCH OF ROCKS

It is believed that humans originated about 200,000 years ago in the Middle Paleolithic period in southern Africa. By 70,000 years ago, humans migrated out of Africa and began colonizing the entire planet. They spread to Eurasia and Oceania, perhaps 40,000 years ago, and reached the Americas by 14,500 years ago.

How long have humans been able to send radio transmissions?

We have only had radio and other types of transmission capabilities for about 100 years plus. Commercial radio transmissions have been around for maybe 90 years (starting in about 1921).

So, while some might dispute the exact time table of development of the Universe, the Earth and man, it has taken place over a *long time* with us only recently being able to send radio or any other transmissions into space.

So, let's say that we have been able to transmit signals since about **1900 CE (AD)**.

What is a prediction of our future progress on Earth?

I have examined some of the fantastic future developments due to computers and other technological developments.

With the development of Virtual Reality technology, it will be possible to do most of our space travel virtually. This will be necessary because of the huge acceleration forces required to obtain reasonable speeds. It will also be necessary because of new technology - which is possible - and may allow space travel, virtually, to some of the closest locations in other solar systems or galaxies.

CAN WE CONTACT OTHER PEOPLE IN THE UNIVERSE?

For example, it might (hopefully) be possible to exceed the speed of light or even "jump" into different space/time "areas." Whatever the new technology, it will probably be desirable to experimentally do our travel virtually so as to save a bunch of lives in the trials.

As pointed out previously, virtual travel will be about as good and desirable to the traveler as real travel. When you consider the safety factor, it will even surpass what we would call superior.

While it isn't possible to accurately forecast the exact developments, they will be stupendous.

Think of the developments in the past 50, 75, 100 years and look ahead to the next 50, 75, 100, 150, 200, 250, 300 years and beyond.

You won't believe it!

So, what will it be like 300 years from now?

We won't be sending many radio signals of the ordinary type. We will be transmitting in coded sequences - performed entirely by computer, of course. We will hopefully be sending signals using new methods of transmitting to exceed the speed of light limitation. Perhaps using quarks or energy strings or who knows what.

We may even have new, faster than the speed of light transmission methods, using different transmission means.

Certainly, we will be able to easily detect any ordinary radio (electromagnetic wave) signals from other places in the Universe. Therefore, we will certainly know if there were/are other people, or beings of a similar nature in the Universe. Then, if we want to, we could contact them. It should be child's play to contact them, at least virtually. At this point, however, be careful to remember that it will probably be many thousands (or billions) of years after the signals were sent that we receive them.

THE UNIVERSE ISN'T JUST A BUNCH OF ROCKS

If we found other beings in the Universe, we could, possibly, visit them. However, if we had a brain in our heads (and we should have computers with the equivalent of many brains), we wouldn't visit them in person. We would, at least at first, take the more prudent step of sending our machines to visit and we could then visit them virtually all that we wanted to. With NO RISK to our bodies, I might add.

Now, after you have tried very hard to visualize the amazing technical progress that we have made in the past 100 years, go for another 100, 200, 300 years. I can't even begin to imagine the future this far ahead. It absolutely boggles my mind, and I've done some boggling in my time too.

So, go back 100 years to when we could first send out radio signals. Then go forward no more than 300 years to where we will be, technically.

This is a 400-year window.

Do you see where I'm heading with all of this?

CAN WE CONTACT OTHER PEOPLE IN THE UNIVERSE?

The 400-year window mismatch

I have pointed out that the evolutionary development of other, potential groups of beings or races will span billions of years. The nearest star to Earth, in the next solar system is more than four light-years away. Any possible planets in the Universe will be, **at least**, a long, long way away from Earth. If we are trying to receive signals from another race, we are looking for, and must find a group that has, in their evolutionary development, this 400-year window that happens to coincide with ours in time.

That is, their evolutionary development must have their 400-year window fit with our 400-year window. Otherwise, in our time, they will either be prior to the development point where they can even send radio signals out or they will be so far developed (past our 300 years in the future development point) - they will probably not be sending ordinary radio signals any more.

To say that in a more correct way, they won't be sending signals anymore. It was many thousands of years ago when they sent the ones that we would be just getting now.

What if the "Aliens" can exceed the speed of light limitation?

First of all, when I say that someone can exceed the speed of light, I really mean it in a little different way than exceeding the limitation directly. So far, our knowledge of technology indicates that this basic limitation seems to be directly tied to all of the equations of space/time/matter/energy.

As previously mentioned: Probably, the way of "exceeding the speed of light limitation" would mean determining how to jump across regions of space.

THE UNIVERSE ISN'T JUST A BUNCH OF ROCKS

If the humankinds from another planet in the Universe *have* figured out a way to exceed the speed of light, then it's a new ball game.

First, if aliens have been able to figure out how to transmit communications signals at faster than the speed of light, they will undoubtedly be using different methods. It is unlikely that we will receive these signals using our normal methods of signal detection.

Also, it's a good guess that they have pretty advanced capabilities in other areas. To start with, I could guess that they are at least a thousand years ahead of us in their technical development. They could be less but the probabilities are that they will be significantly way ahead in most areas.

Their "window" of development is probably thousands or millions of years out of line with our present 400-year window.

Now, also consider the possibility that the aliens might want to visit us. They will be so advanced that visiting us is virtually child's play. Remember, since they are many, many years of development ahead of us, their "space ships" will be so far advanced as to be impossible to detect, if they didn't want to be detected. Making a "space ship" invisible will be really, really easy in even just 100 to 200 years. In thousands of years it will be total "duck soup" with computer technology being capable of redirecting every type of light and other kind of wave used to detect the vehicle's presence.

CAN WE CONTACT OTHER PEOPLE IN THE UNIVERSE?

It would be entirely up to the aliens whether they wanted to be detected or known, not up to some "old-technology flying saucer" stuff. Probably the most obvious realization that I can have about visiting aliens (if they do exist), is that they certainly would want, for safety, to stay several thousands of miles away from Earth while their probes examined our planet. They would use their Virtual World to visit us in complete safety.

THE UNIVERSE ISN'T JUST A BUNCH OF ROCKS

18   The Universe - The Future of Humans

It is important to project the evolution of the lifetime span of humans as well as the evolution of computers and technology in order to begin to get a better idea of the future "big picture" of our human lifestyle.

Life expectancy of humans

Life expectancy of humans is the average number of years a person can expect to live, usually noted as the expected length of life at the time that the person is born.

The dramatic increase of life expectancy in the past 150 years is due to advances in public health, nutrition and medicine. Of course, life expectancy varies by country, sex and ethnicity.

The following are some composite figures of the Life Expectancy at Birth in the U.S.:

Year	Age
1850	38
1900	49
1930	59
1960	70
2000	76
2006	77
2025	79
2050	81
2075	83
2300	102

Lifespan of humans

Lifespan is different than life expectancy.

Lifespan is the maximum number of years a person lives if all natural diseases and accidents are optimum. The current oldest person (verifiable) on record in the world is 122 years old and there are very, very few older than 110. There are many unverified stories of older people throughout history, of course.

Maximum lifespan has not changed appreciably since reliable records were maintained.

This all means that the next step in the extension of Life Expectancy must be to solve some of the basic aging processes which limit it to less than 122 years.

Extending the lifespan of humans

Cambridge University geneticist Aubrey de Grey has famously stated, "The first person to live to be 1,000 years old is certainly alive today ....whether they realize it or not, barring accidents and suicide, most people now 40 years or younger can expect to live for centuries."

Perhaps de Gray is way too optimistic, but plenty of others have joined the search for a virtual fountain of youth. In fact, a growing number of scientists, doctors, geneticists and nanotech experts - many with impeccable academic credentials - are insisting that there is no hard reason why aging can't be dramatically slowed or prevented altogether. Not only is it theoretically possible, they argue, but it is a scientifically achievable goal that can and should be reached in time to benefit those alive today.

Even the U.S. government finds the field sufficiently promising to fund some of the research. Federal funding for "the biology of aging," excluding work on aging-specific diseases like heart failure and cancer - has been running at about $2.4-billion a year, according to the National Institute of Aging, part of the National Institutes of Health.

So far, the most intriguing results have been spawned by the genetics labs of bigger universities, where anti-aging scientists have found ways to extend life spans of a range of organisms - including mammals. But genetic research is not the only field that may hold the key to eternity.

"There are many, many different components of aging and we are chipping away at all of them," said Robert Freitas at the Institute for Molecular Manufacturing, a nonprofit, nanotech group in Palo Alto, California. "It will take time and, if you put it in terms of the big developments of modern technology, say the telephone; we are still about 10 years off from Alexander Graham Bell shouting to his assistant through that first device. Still, in the near future, say the next two to four decades, the disease of aging will be cured."

But not everyone thinks aging can or should be cured. Some say that humans weren't meant to live forever, regardless of whether or not we actually can.

"I just don't think [immortality] is possible," says Sherwin Nuland, a professor of surgery at the Yale School of Medicine. "Aubrey and the others who talk of greatly extending lifespan are oversimplifying the science and just don't understand the magnitude of the task. His plan will not succeed. Were it to do so, it would undermine what it means to be human."

After all, we already have overpopulation, global warming, limited resources and other issues to deal with, so why compound the problem by adding immortality into the mix.

Whatever your view of the immediacy of making large strides in solving the aging problem, it seems reasonable to assume that there will be great improvements in the next several decades.

Add the control of diseases and artificial or regenerative organs and it seems almost certain that lifespan will be significantly increased with a comparable increase of life expectancy.

Let's conservatively assume that the average life expectancy in the

THE UNIVERSE ISN'T JUST A BUNCH OF ROCKS

U.S. in the year 2112 (100 years from now) will be 140 and will be 200 in the year 2212 (200 years from now). These numbers may be too conservative and we may be able to better them considerably.

Can you live forever?

One interesting idea in considering the extending of the lifespan of humans is shown in one of my favorite (over simplified) conjectural stories:

If you can extend the lifespan by 50 years in the next 100 years, then you will live through the next extension of lifespan. Then, if you keep living through the next lifespan extension, the lifespan will again be extended, and on and on.

Following this line of reasoning: **If you can live for another 100 years, you can live forever**.

Isn't this exciting? Remember that I said it was an over-simplified story!

19 Our Responsibility to Limit Population, in the Future

World Population Growth

The total number of people that have ever lived is estimated at 110 billion. A rough idea of the world population growth is:

Time Period	World Population
10,000 BC	1,000,000
5,000 BC	5,000,000
1,000 BC	50,000,000
0 AD	200,000,000
500 AD	300,000,000
1000 AD	400,000,000
1500 AD	500,000,000
1800 AD	900,000,000
1850 AD	1,200,000,000
1900 AD	1,600,000,000
1950 AD	2,400,000,000
1970 AD	3,700,000,000
2000 AD	6,100,000,000
2012 AD	7,000,000,000
2020 AD	7,600,000,000
2030 AD	8,200,000,000
2040 AD	8,800,000,000
2050 AD	10,000,000,000

THE UNIVERSE ISN'T JUST A BUNCH OF ROCKS

Where are we heading and what must we do?

An examination of evolutionary development of the reproductive process:

The system for reproducing all forms of life, including humans, has worked for a long time.

It may be theorized that life would have died out long ago without the basic built-in mechanisms for continual reproduction of life.

It is also interesting to theorize as to how the development of reproduction was supported by the theory of natural selection. It must have taken a lot of chance developments of life without *reproduction* before the natural-selection process selected reproduction as a way to keep things going. In other words, without reproduction, lots of new life forms had to continuously keep coming about through chance before life got to be self-perpetuating.

Boy! If I were in charge of evolution in its early days, I would have gotten pretty discouraged before "chance" finally got around to trying the reproducing thing. I would have had to keep starting life from nothing over and over and over again, from scratch without any automatic reproduction.

When *Chance*, *Survival of the Fittest* and *Natural Selection* finally came up with sex, I would have breathed a monster sigh of relief.

Sometimes, I wonder if there could have been a hand of something like a creator in this to suggest to nature that it might be a good idea for reproduction to be an absolute part of all of life - to keep it going without life having to keep starting from the ground up so darn many times.

Again, the idea of a Planner and Creator really makes some sense.

At any rate, reproduction is part of the life process and it has worked fantastically well for a long, long time - and then some.

OUR RESPONSIBILITY TO LIMIT POPULATION, IN THE FUTURE

A look at overpopulation of animals

In some cases, as with lower forms of life, there seems to have been something in the overall formula that keeps adjusting the service of reproduction so as to keep the numbers of a species from getting completely out of control.

Overpopulation in wild animals occurs when a population of a wild species exceeds the carrying capacity of its ecological niche. Overpopulation is a function of the number of individuals compared to the relevant resources, such as the water and essential nutrients they need to survive.

In the wilderness, the problem of animal overpopulation is solved by predators. Predators tend to look for signs of weakness in their prey, and therefore usually first eat the old or sick animals. This has the side effects of ensuring a strong stock among the survivors, and controlling the population.

In the absence of predators, animal species are bound by the resources they can find in their environment, but this does not necessarily control overpopulation. In fact, an abundant supply of resources can produce a *population boom* that ends up with more individuals than the environment can support. In this case, starvation, thirst and sometimes violent competition for scarce resources may effect a sharp reduction in population in a very short lapse.

Some animal species seem to have a measure of self-control, by which individuals refrain from mating when they find themselves in a crowded environment. This voluntary abstinence may be induced by stress or by pheromones.

A look at overpopulation of people

People seem not to have the advantage of using most of the basic methods that wild animals do to control overpopulation.

THE UNIVERSE ISN'T JUST A BUNCH OF ROCKS

Although there are predators who deal with humans, they just take another person's money or freedom; they don't usually take their victims' lives. If humans get sick in the developed world, they usually go to a doctor, hospital or a free clinic where they receive care.

If people exceed the number of individuals compared to the relevant resources, such as the water and essential nutrients that they need to survive, they just keep on reproducing.

Starvation in third-world countries has caused some significant reduction in populations but, on a global scale, the effect has been minimal.

In the distant past, there were some plagues and wars that did seem to reduce the population. However, largely due to modern technology and such, these can't seem to be counted on to cause a significant reduction in population.

And, people don't seem to have any measure of self-control by which these individuals refrain from mating when they find themselves in a crowded environment.

How bad is the problem?

As you can see from the numbers of population growth previously shown, things are rapidly getting out of control. Actually, it may be argued that they have already gotten out of control - way out!

The number of people on Earth is increasing so fast that I hate to tell you about the next thing that I have to say:

In the preceding section, it was pointed out that the average life expectancy is going up all the time. Also, that some experts are confidently predicting that the maximum life *span* will go up radically in the near future. One expert stated that she expects many people now in their 30s will live to be 1000. Wow! Think of what that will do to the population estimates shown previously.

OUR RESPONSIBILITY TO LIMIT POPULATION, IN THE FUTURE

(One more thought: For years, Jehovah's Witnesses have taught that only a fixed number of people will go to heaven when they die but that the rest of the good people will remain in a paradise on Earth. Now, with the previously shown estimates of population, the earth will get pretty darn crowded and I mean that! However, if people start living to 1000 years and beyond, there won't be room for people's wooden shoes to be placed by their doors in Holland or room for the Japanese people to place their shoes when removing them before entering their homes, either.)

At any rate, people reproduction has worked fantastically well for a long, long time - and then some, but enough is enough.

Obviously, something has to be done.

It's time to put the brakes on this rampant, completely out of control, population growth.

We have to get past this wild upward spiral into a "packed house.

And soon!

What actions do we take?

It appears that the only alternative is to take some global action. We only have two options: Kill people off or inhibit the start-ups. Obviously, the first thing to do is to implement methods which control birth rates.

This is quite a political problem with cultural, religious, and other social implications.

It seems obvious that the first thing to start with is to provide options for all people to use birth-control methods and encourage them to do so.

THE UNIVERSE ISN'T JUST A BUNCH OF ROCKS

I am afraid that I must leave this problem for others to work on. **I only know that it is really imperative, for everyone's health and welfare, to solve the birthrate problem, and fast!**

Then, one final comment:

Religious leaders absolutely must balance the desire to restrict use of any kinds of birth control with a humanitarian attitude.

Doesn't the suffering of thousands and thousands of people matter at all?

20 What About "Miracles"

Definition:

An effect or extraordinary event, in the physical world, that surpasses all known human or natural powers and is ascribed to a supernatural cause.

An extraordinary event which manifests divine intervention in human affairs. An extremely outstanding or unusual event, thing, or accomplishment.

A wonder; marvel.

A further, more precise definition

What it really boils down to is that miracles are usually one of two types:

1. *An event that happens without really violating any laws of nature, but which has a very low probability of occurrence. Perhaps with divine assistance.*

2. *An event which violates, or supersedes one or more rule or law of the natural world. This type can only be, it is assumed, performed by divine intervention since, assumedly, the rules of the Universe are just that - rules - and can't, by themselves, be violated.*

3. *An event which seems to violate or overcome a natural law, but which is really just advanced technology "getting around" the natural laws. It must be said that this advanced technology could be from a divine source.*

Miracles, a discussion

I think that people's prayers for assistance might get answered, but I get the feeling that the prayer answerer would probably use his "Universe rules" and not violate them in helping people.

THE UNIVERSE ISN'T JUST A BUNCH OF ROCKS

All indications are that God, as the creator of the Universe with the myriad of laws and rules that he must have created, doesn't strike me as the kind who would then violate these rules. Of course, he might have a way of accomplishing many, many fantastic things by using the "rules" that he created through technological advances - perhaps known only to him.

One of the characteristics of these "rules" or "laws of nature" is that they are absolute. This is why they are called rules. It just doesn't make any sense that they could be violated or they aren't really rules/laws.

No, rules (laws) would seem to be just that - laws.

Why would the creator set up the entire myriad of interacting, self-adjusting, coordinated rules and then violate them - if they could be violated.

Also, he wouldn't need to. He could accomplish things by using scientific technology - which he certainly would have FULL access to. For, example, if you want to fly, then you might use an airplane - or something better. If you want to cure somebody of a disease, use medical techniques or psychology to cure them. It would make sense that advanced technology, while following the rules/laws, would allow the effect of overcoming these rules/laws.

Advanced computer capabilities of one-hundred to several hundred years from now will allow fantastic accomplishments which will seem like miracles. (Consider that many of the commonplace technological advances at the present time would seem like real miracles to people a few years ago!)

In summary, it would seem that miracles where the real rules are just *magically* overcome and violated - do not make much sense at all.

WHAT ABOUT MIRACLES?

Perhaps, I should be extremely careful in describing "God" in the terms of humans and what I would like to see, instead of what the creator - God, might do.

Think about it. Please, just think about it.

God's rules/laws about the physical Universe

We have talked about the many, interwoven rules of the Universe.

Would these physical rules/laws have been set up and then violated often?

Thus, the law of gravity, Sir Isaac Newton's laws of motion, the limitations of the speed of light, etc. are here to stay and not to be treated lightly or violated. This is true even in answer to anyone's prayers to God for help. It is even true when a husband who has stayed out too late asks for help in confronting the little woman, when returning at 2 a.m. after a fun night out with the "boys."

God, as the creator, set things up to run pretty much by themselves which they do quite well. They do this greatly aided by the basic laws of physics, which seem to govern the Universe unconditionally.

I ask you, "Why would God set up all of these laws and rules of physics and then allow them to be violated, even in answer to prayers by his friends?"

If I were God (I know, a really presumptuous assumption), and I wanted to make something occur, I would not try to violate my rules, even if I could. For one thing, this is a *contradiction* of nature. For another, it could cause serious repercussions with all the other rules/laws of nature.

THE UNIVERSE ISN'T JUST A BUNCH OF ROCKS

Another reason that I wouldn't have to, is that I wouldn't *need* to. If God created the Universe (and I have shown that there was a creator), surely, he could come up with some new technological whiz-bang idea that would let him do whatever he wanted to - without violating his *inviolate* laws.

Wouldn't that make sense?

OK, then, how about records of miracles?

Some other accounts of miracles can be explained in reasonable ways.

Some time ago, I had a dream where I suddenly found that I could walk on the surface of water by "shuffling my feet" (typical dream sequence type). I was telling other people how to do it and they were walking on the surface of water also. When I awoke, I, of course, found that that didn't work in actual practice. Then, I got to thinking about the account of Jesus walking on water late at night to get out to the boat where his disciples were fishing. The description made me think that, possibly, the account coming from a disciple on the boat could have been his account of a late night, drowsy dream. It was quite similar to my own dream. This is a possibility.

A reasonable conclusion is that there must be some explanation for most miracles that leave God's rules (laws) intact and, basically, unchanged.

Perhaps, many reasonably verified miracles can be explained by God adjusting the laws of probabilities.

WHAT ABOUT MIRACLES?

Adjusting the Laws of Probability to perform miracles

The Mathematical Laws of Probability seem reasonable to mathematicians and to regular people. Mathematical formulas have been written about these laws and expounded on in thousands of books, papers, and speeches. Certainly, there is adequate confirmation by observing things in the world. Of course, Las Vegas is a solid example of that.

We just know that we wouldn't be too worried if we were before a firing squad with nine other people. And, in this situation, there were 10 soldiers, each aiming at one of us. And ONLY ONE OF THE SOLDIERS had a bullet in his gun. Well, maybe, we'd be a little worried, but you get the idea.

Probabilities seem to really work solidly throughout the world. The mathematical equations work in everyday life. However, these mathematic formulas (rules) have never been actually *proven* by logical physical sequences. They are mostly proven by logic that assumes probabilities when multiple occurrences are equally or proportionately likely. This is to say that, if there are several possibilities and there is no apparent reason that any one of them should have any better reason for occurring than any of the others, that they would have equal probabilities. This means that each side of a six-sided die (one of two dice) with six different dot combinations would have an equal probability of coming up when the die are rolled. This treatment of probabilities has, of course, been verified by observing that the formulas work in practice.

However, if probabilities were violated only once in a while, and only under special conditions in answer to a prayer, this wouldn't cause the world to go into total chaos. By the way, I think that gamblers' prayers would be seldom answered - so, Atlantic City, Reno, and Las Vegas, relax.

Perhaps, God or his special representative, or the "rules of the Universe," in answer to prayers, might adjust the probabilities of a situation to provide the solution.

The conclusion about miracles

It is my strong feeling that God, the Creator of the Universe and of all integrated laws would NOT try to violate these laws directly. In fact, as previously mentioned, that is actually a contradiction (to set up laws and then to violate them).

The idea of miracles, where these are actual violations of the laws themselves, seems to be the exact type of thing that people would make up in order to promulgate fanciful stories of *magic*.

It is my further reasoning that the Creator would use advanced technology and, perhaps, adjusting of probabilities to accomplish his goals in the *miracle* department.

I do believe that reason points us to the conclusion that the Creator *does* do things of this nature in order to help his goals and, in order to help people in response to their requests.

21 About Heaven

The idea of *heaven* has existed for many years and is embraced by many religions. Many religions state that those who do not go to heaven will go to a place called hell. Often there are conditions for entry to heaven. For example, it may be a reward for good works.

The idea of heaven is sharply argued by many Jews. The idea of heaven is not mentioned in the Christian gospel of Mark - perhaps the earliest of the gospels. Also, it is not mentioned in the letters of Paul. It has been suggested that the concept of heaven in Christianity may have superseded the often mentioned "Kingdom of God" (on Earth) mentioned in Mark.

Many religions also consider a future Judgment Day where each soul is judged.

At any rate, when I consider the concept of heaven or of an afterlife, I must conjecture since I have no actual proof of its existence.

A good friend of mine used to remark, "Well, nobody ever came back."

So, we can accept one or several of the concepts expounded by religions or come up with our own ideas but we just don't know for sure.

What I can do is to present some basic *Heaven Idea* facts which are food for thought.

Afterlife in Heaven

We could go to an afterlife in heaven. Under this concept, the good would go there and the rest might go to some version of a hell.

The most reasonable idea of this heaven would place it as a non-physical existence. The Universe started and therefore had to be created and must have had a creative force called God. God must have existed in a non-physical dimension without time/space/energy/matter *prior* to this creation.

It would seem to make some sense to assume that heaven would be in this domain. This concept would seem to be supported by reason since, otherwise, where would it be? There are a number of religious concepts of an afterlife where we might keep our physical images but this seems to be hard to explain very well in detail.

Let's, therefore, assume that this heaven (and hell, also, apparently) will have no physical characteristics such as time/space/energy/matter.

The primary conclusion that I must make is that I can't really describe this heaven at all. However, it does sound preferable to simply terminating our existence upon death.

We just go poof

Even though I don't like this idea, it does seem to be the most logical of the possibilities.

There are a number of interesting possibilities when considering this concept. For example, could our DNA be saved and our bodies recreated in the future? Could our bodies be frozen and defrosted in the future when technology would be vastly improved? Then, what about our memories? What about our souls?

Under this concept, our heaven or hell might involve the success of our descendants after our influences on them. Our afterlives might be in those of our descendants or those who we influenced.

ABOUT "HEAVEN"

Reincarnation

Reincarnation describes the concept where the soul or spirit, after the death of the body, is believed to return to live in a new human body, or, in some traditions, either as a human being, animal, or plant.

If a person believes that good is rewarded by some future existence in a place such as heaven, the idea of reincarnation is absolutely logical. This concept overcomes the tremendous inequities of a single lifespan on Earth. Why should a person who lives for only two days after birth be judged the same as a person who lives for 80 years? A person who lives only in an incubator can hardly have the same chance for "good advancement" as a person with 6 children and 15 grandchildren. What about a person who is born blind and deaf?

The idea of multiple lives for a soul resolves many of these inequities on the final *judgment* day.

Personally, I would like to be given multiple lives as a person rather than spending too many cycles as a daffodil or an amoeba.

THE UNIVERSE ISN'T JUST A BUNCH OF ROCKS

22 A Simple Conclusion

When we gaze at the sky at night, we see many multitudes of stars in all directions. Scientists tell us that there are actually many billions of these out there. Most of these we can't see with the naked eye or even with the most powerful telescopes.

When I think of the vastness of the Universe and the billions of all kinds of bodies, I think of the matter of all kinds, the gases, the energy with super novas exploding - and on and on.

I envision the vastness of the distances between galaxies and the time that it takes light and other energies to go from one area to another.

Remember that these things pale into insignificance when compared to the rules/laws that really make up the most important elements of the Universe. It's these rules that really hold together all of the pieces of the Universe, determine how the electrons attract, how the atoms clump to form molecules to make up elements of matter, etc.

Time and space, together with the energy and the matter involve equations which spell out everything working together in distinct harmony.

While evolution is undoubtedly involved in the development of many life forms, it seems impossible to consider that the RULES/LAWS of the Universe could have developed from nothing, using accidental *trial and error* and *survival of the fittest*.

Somehow, there must have been a *guiding hand* in the development of these RULES.

THE UNIVERSE ISN'T JUST A BUNCH OF ROCKS

ACKNOWLEDGEMENTS

I would first like to thank God for his help to me in formulating my thoughts for this book.

I really appreciate my mother and father, Minnie and H. Edgar Williams (now deceased). I was fortunate to have these absolutely outstanding people show me and my siblings the real importance of religion and adherence to God's principles in the treatment of everybody. Dad never swerved from his adherence to God's goals and he also taught me that remaining an individual and questioning was an important part of it all.

My wife, Rosemary, who died two years ago, had been the most loyal, intelligent, and important person to me for many, many years. She lived a life as close to what Jesus suggested as anybody who I know, always helping everyone. She was born, and lived, a Catholic life; but she absolutely insisted on not judging and on allowing everyone to follow his or her own ideas and faith, even if that faith wasn't Catholic.

I would like to thank Carol Williams for all of her help in advising me, pointing out when I sounded "less than correct," and in editing this book. Her independence is an important part of her character.

I appreciate my sister, Nancy Blair, and her husband, Glenn, for their constant encouragement and enthusiasm toward this project. I appreciate the friendship of Glenn and his unwavering support over many years.

To my brother and sister-in-law, John and Carol Williams, and my sister and brother-in-law, Margaret and Kenneth Carpenter - thanks for your friendship and examples in life.

THE UNIVERSE ISN'T JUST A BUNCH OF ROCKS

I appreciate all of Rosemary's and my wonderful children in living really good, creative, and productive lives. To Andrew and Amy Williams, Luci Hummel, Kathryn Carter, Mary and Gary Miller and to Charles and Beth Williams - thank you for your outstanding adherence to good principles. Also, thanks for not ever "going to the dogs." I appreciate it. (Kate and Mark Peluso, thanks for your advice on this book.)

To all of my fine 13 grandchildren and spouse, three great grandchildren, and to my grand and great nieces and nephews, I salute you also.

Thanks to my nephew, Rick Blair, for his technical advice in areas of the book.

I also owe a statement of appreciation to all of the people with whom I worked for many years in developing and improving our Hospital Information System. Many of these people gave much more than just "doing their job" in our work, working many long hours, unrewarded, in the early days. The examples of many of these people are also a testimony to the value of "goodness" in our lives.

I'll name just a few of the many involved with the development of this fantastic Hospital Information System. To Dr. Brad Hisey, who had the original idea of the "Video Matrix" concept used for data entry. To my friend, Chuck Tapella, for his remarkable analysis and "impossible" programming work when he would do what nobody else could. To Dave Brown, Bob McCord, Gary Lager and Ralph Boyce for their outstanding technical ability and for their friendship. To Bill W. Childs for his support and friendship through the years after he left the company and became a nationally known and respected health-care consultant - and who made a constant contribution to the less fortunate. To Melville Hodge and Kenneth Larkin for their corporate management support (they saved our company's life) throughout the first years of our struggle before things got easier. And to Jack Whitehead, owner of Technicon Corporation, who owned us and really believed in us in the early years.

And to the many, many others who had an important part in working with this system, too numerous to name - thanks for your example.

I really appreciate the many, many people who I have met and worked with at a number of churches - especially to the 10-to-20 "6:30 daily mass regulars" at Ascension Church in Saratoga, California. These, people were, and are, a constant reminder of what religion is *all about*.

A special note of thank you for support to Nate Bernstein, Tillie Goldsberry and to Paul and Phyllis Thomas.

Of course, there are the many other friends and acquaintances, of various religious faiths who have always inspired me with their examples and with their friendships.

THE UNIVERSE ISN'T JUST A BUNCH OF ROCKS

About the author

Paul E. Williams was married for 54 years, with five children, 13 grandchildren and three great-grandchildren. His wife, Rosemary, was a wonderful partner, mother and grandmother. Unfortunately, about 13 years ago, she started developing Alzheimer's and died in January, 2010.

Starting in the late 1960s, Paul E. Williams was the chief architect of and directed the development of a hospital information system. This comprehensive system used CRT terminals with light pens throughout the hospital for doctors to enter their orders, nurses to chart patients' records, medication personnel to chart medications, and all other departments to assist in their duties of patient care. This system was first called the "MIS System" and later, the "TDS Healthcare System." It was first introduced at El Camino Hospital in Mountain View, California in 1970. At that time, it was very early for hospital professionals to use computers directly - and very expensive in relation to other hospital costs and prices. Eventually, the system came into its own as other costs went up and was used in 350 hospitals in the U.S. and Great Britain, including Indiana Methodist in Indianapolis (at 1200 beds, the largest), and National Institutes of Health Hospital in Bethesda, Maryland (perhaps the most prestigious). This system is still used at approximately 30-50 hospitals, even though it is 40-years old.

Books published by Paul E. Williams include "If I Had Been Born a Muslim,, 2010. Paul E. Williams graduated from the University of Denver in 1952 with a major in mathematics and a minor in philosophy.

THE UNIVERSE ISN'T JUST A BUNCH OF ROCKS

THE UNIVERSE ISN'T JUST A BUNCH OF ROCKS

THE UNIVERSE ISN'T JUST A BUNCH OF ROCKS

THE UNIVERSE ISN'T JUST A BUNCH OF ROCKS

THE UNIVERSE ISN'T JUST A BUNCH OF ROCKS

www.ingramcontent.com/pod-product-compliance
Lightning Source LLC
Chambersburg PA
CBHW030927180526
45163CB00002B/489